Research Series on the Chinese Dream and China's Development Path

Drawing on a large body of empirical studies done over the last two decades, this Series provides its readers with in-depth analyses of the past and present and forecasts for the future course of China's development. It contains the latest research results made by members of the Chinese Academy of Social Sciences. This series is an invaluable companion to every researcher who is trying to gain a deeper understanding of the development model, path and experience unique to China. Thanks to the adoption of Socialism with Chinese characteristics, and the implementation of comprehensive reform and opening-up, China has made tremendous achievements in areas such as political reform, economic development, and social construction, and is making great strides towards the realization of the Chinese dream of national rejuvenation. In addition to presenting a detailed account of many of these achievements, the authors also discuss what lessons other countries can learn from China's experience.

More information about this series at https://link.springer.com/bookseries/13571

Peilin Li
Editor

The Quality of Life in Contemporary China

SSAP

Springer

Editor
Peilin Li
Chinese Academy of Social Sciences
Beijing, China

ISSN 2363-6866 ISSN 2363-6874 (electronic)
Research Series on the Chinese Dream and China's Development Path
ISBN 978-981-16-7020-6 ISBN 978-981-16-7021-3 (eBook)
https://doi.org/10.1007/978-981-16-7021-3

Jointly published with Social Sciences Academic Press
The print edition is not for sale in China (Mainland). Customers from China (Mainland) please order the print book from: Social Sciences Academic Press.

Translation from the Chinese language edition: *The Quality of Life in Contemporary China* by Peilin Li, © Social Sciences Academic Press 2018. Published by Social Sciences Academic Press. All Rights Reserved.

This Springer imprint is published by the registered company Springer Nature Singapore Pte Ltd.
The registered company address is: 152 Beach Road, #21-01/04 Gateway East, Singapore 189721, Singapore

Series Preface

Since China's reform and opening began in 1978, the country has come a long way on the path of socialism with Chinese characteristics, under the leadership of the Communist Party of China. Over 30 years of reform, efforts and sustained spectacular economic growth have turned China into the world's second largest economy and wrought many profound changes in the Chinese society. These historically significant developments have been garnering increasing attention from scholars, governments, and the general public alike around the world since the 1990s, when the newest wave of China studies began to gather steam. Some of the hottest topics have included the so-called China miracle, Chinese phenomenon, Chinese experience, Chinese path, and the Chinese model. Homegrown researchers have soon followed suit. Already hugely productive, this vibrant field is putting out a large number of books each year, with Social Sciences Academic Press alone having published hundreds of titles on a wide range of subjects.

Because most of these books have been written and published in Chinese, however, readership has been limited outside China—even among many who study China—for whom English is still the lingua franca. This language barrier has been an impediment to efforts by academia, business communities, and policy-makers in other countries to form a thorough understanding of contemporary China, of what is distinct about China's past and present may mean not only for her future but also for the future of the world. The need to remove such an impediment is both real and urgent, and the *Research Series on the Chinese Dream and China's Development Path* is my answer to the call.

This series features some of the most notable achievements from the last 20 years by scholars in China in a variety of research topics related to reform and opening. They include both theoretical explorations and empirical studies and cover economy, society, politics, law, culture, and ecology, the six areas in which reform and opening policies have had the deepest impact and farthest-reaching consequences for the country. Authors for the series have also tried to articulate their visions of the "Chinese Dream" and how the country can realize it in these fields and beyond.

All of the editors and authors for the *Research Series on the Chinese Dream and China's Development Path* are both longtime students of reform and opening and recognized authorities in their respective academic fields. Their credentials and expertise lend credibility to these books, each of which having been subject to a rigorous peer review process for inclusion in the series. As part of the Reform and Development Program under the State Administration of Press, Publication, Radio, Film, and Television of the People's Republic of China, the series is published by Springer, a Germany-based academic publisher of international repute, and distributed overseas. I am confident that it will help fill a lacuna in studies of China in the era of reform and opening.

Shouguang Xie

Introduction: Towards a Greater Emphasis on the Quality of Life

China has set itself the goal of building a *xiaokang* society in all respects by 2020. *Xiaokang* is a traditional Chinese expression for what may be understood as a "comfortably well-off life." Before being used to symbolize a milestone in the course of China's modernization, the phrase traditionally described a sort of peaceful, idyllic village life. In this book, we shall refer to *xiaokang* society by its official English translation: "moderately prosperous society,"[1] or more comprehensively, "moderately prosperous society in all respects."

"Moderately prosperous" refers more to the quality of life (as measured by comprehensive social indicators) rather than merely the level of economic development (as measured by per capita GDP). After 38 years [*sic*[2]] of reform and opening up, China has seen rapid economic development, momentous social change, and a steady rise in the living standards of her people. These same years have also seen China moving toward a greater emphasis on the quality of life.

From the "Take-Off Mode" to a New Stage of Growth

In 2008, the 30th anniversary of reform and opening-up, GDP per capita in China exceeded USD 3000 for the first time. Various attributes of China's development during that period led me to believe that China had in fact begun to embark upon a new phase of socioeconomic development. The fundamental characteristics of socioeconomic development during that period were, to a very large extent, completely unrecognizable from that of the "economic take-off mode" three decades ago. In

[1] See, for instance, the official work report delivered by Xi Jinping at the 19th CPC Congress.—Trans.

[2] This book was originally published in 2016. China launched economic reforms known as "reform and opening up" (*gaige kaifang*) in 1978, which made 2016 the 38th anniversary of China's economic reforms.—Trans.

other words, I was arguing that China had by then already moved past the take-off mode.

In that case, what stage of development did China enter after her economy had taken off? Of course, according to the development goals that China had set for herself, we may say that the country had entered the stage of "building a moderately prosperous society in all respects," which is an overarching term for the stages of development that would span from 2000 to 2020. However, this characterization does not lend itself easily to international comparison. Put differently, how is the stage of "building a moderately prosperous society in all respects" related to the universally accepted concept of an "(economic) take-off mode"?

The idea of an "economic take-off" was first proposed by the American economist W. W. Rostow in his book, *The Stages of Economic Growth*, published in 1960. Rostow divided the economic growth of a country and society into five stages (later increased to six[3]): the traditional society, the preconditions for take-off, the take-off, the drive to maturity, the age of high mass-consumption, and "beyond mass-consumption."

However, most of Rostow's writings on these stages were merely descriptive, with the exception of the "take-off mode." His research focused on the take-off mode, for which he provided a more detailed definition. Rostow argued that, in "taking off," an economy overcomes traditional economic structures, for which three preconditions are required. First, there must be a rise in the rate of effective investment and savings to 10% of national income or more. Second, take-off requires the predominance of a modern industrial sector. Third, there must be institutional changes, i.e., the establishment of political, economic, and social institutions that allow for the expansion of the modern sector. Among the Western countries, Britain "took off" during the last two decades of the eighteenth century, France and the US "took off" over some decades before 1860, while Germany achieved the same between 1850 and 1875. Similarly, Japan achieved her take off in the last 25 years of the ninteenth century.

Thus, we see that these countries took roughly two to three decades to complete their economic take-offs. From these dates, it is also apparent that Rostow's conception of the economic take-off was largely equivalent to basic industrialization. Rostow also proposed certain specific indicators characteristic of an economic take-off: his theories in this respect would later be known as the Rostovian take-off mode. However, the other stages that Rostow spoke of have mostly been forgotten. The age of high mass-consumption, frequently mentioned in the media, is perhaps an exception, although it has so far failed to become an established theory.

After reform and opening-up, I was one of the earliest researchers to characterize the course of China's long-term development as a "social transformation." By social transformation, I meant China's transformation from an agricultural, rural

[3] Rostow would go on to describe a sixth stage in his 1971 book *Politics and the Stages of Growth*, which the author termed the "beyond mass-consumption stage" (*chaoyue dazhong xiaofei jieduan*). Rostow himself originally called this stage "the search for quality." See W.W. Rostow, *Politics and the Stages of Growth* (Cambridge: Cambridge University Press, 1971), p. 22.—Trans.

and (semi-) closed society to an industrial, urban, and open society. In other words, the usual process of modernization experienced by countries and societies, marked by industrialization and urbanization. This is a theoretical generalization based on common knowledge. Yet, back when most attention was focused on the transition from a planned economy toward a market economy, this characterization provided a perspective from which to scrutinize development from a longer horizon and greater dimensions, which assumed that momentous social changes would continue even after the economic system had been reformed. The transformation of the social structure would become a different force that drives social change—a force that is more fundamental and long-term in nature.

This characterization of China's development from "pre-traditional to modern," which encapsulated a rather long horizon, was not without its drawbacks: It does not clearly define the various stages in between, which makes it difficult to form a workable definition that may be directly applied to the analysis of each specific stage. Hence, we require a new characterization of the post-take-off stage, which would allow us to analyze the changes that have taken place during this phase in China's development. Ultimately, I decided on the concept of a "new growth stage."

In late 2009, when I was writing the general report for *Social Analysis and Forecast of China 2010*, I opted for the heading "A New Growth Stage of Development." At the time, the report outlined the characteristics of this new growth stage from six aspects:

1. Industrialization and Urbanization Advancement Enters Middle Period of the New Growth Stage
2. Changes in Social Structure Breaks with the Dualistic Structure of Urban and Rural Areas
3. People's Lifestyle Enters a New Growth Stage of Mass Consumption
4. Higher Education Enters New Growth Stage of Mass Education
5. Social Security Enters New Growth Stage of Extensive Coverage
6. From Economic Reform to Comprehensive Reform, Transition to a New Growth Stage

Thereafter, I made repeated attempts to formulate a more persuasive characterization of the new growth stage to make clear that this is a new stage of development radically different from the past. In late 2014, I wrote a Preface for *Social Analysis and Forecast of China 2015* titled "The New Stage of Growth Against the Backdrop of the 'New Normal'." In this preface, I outlined the following major attributes of China's current transition between different stages of development.

First, the transition between different stages of urbanization. At the global level, urbanization has generally been sub-divided into four phases: (1) urban population growth, (2) suburbanization, (3) de-urbanization, also known as counter-urbanization, and (4) re-urbanization. China's level of urbanization first exceeded 50% in 2011 and has reached 56.1% in 2015. Urbanization in China can be described as a form of accelerated, leapfrog development. On the one hand, the overall urban population is still growing. On the other hand, China is already seeing trends of suburbanization and de-urbanization. This indicates the dawn of a new stage: Though these trends are as yet largely manifested in the forms of "agrileisure," retirement in rural villages, and the running of tertiary industries in rural areas by urban residents, they are signs of a future trend. Suburbanization and de-urbanization do not imply a reversal of urbanization, but rather a new phase of urbanization with greater urban–rural integration that will breed vast new opportunities for further development. Although China's huge population makes it unlikely that urbanization can be achieved merely through population influx into cities, it is nevertheless likely that urbanization levels in China will continue to increase until sometime around 2035, when 75% of the Chinese population will be living in urban areas. In this respect, China still has great flexibility for structural adaptation and room for development, making it advisable to move along with prevailing trends and actively promote new forms of urbanization. These new trends will also require China to find new ways to manage internal migration. Given the urban–rural stratification in the household registration (*hukou*) system, it will be necessary to integrate the hundreds of millions of rural migrants flowing into cities, as well as manage the flow of urban residents into rural areas, whether for leisure or business.

Second, shifts in the labor market as China transitions between different stages of development. Unemployment rates in China have not increased despite the slowing economic growth rates—urban unemployment surveys, which reflect the actual rates of unemployment, have been closely aligned with registered urban unemployment rates (which are somewhat statistically limited in terms of sample size). Given that previous slowdowns had led to higher unemployment, this was a rather pleasant surprise. Despite slowing economic growth, companies in the Yangtze Delta and Pearl River Delta were instead worried that migrant workers would not return after the Spring Festival break—which had not been the case during the cyclical fluctuations of the past decades. This was mainly due to three factors:

1. Immense government efforts to promote job creation, which resulted in more than ten million new jobs in 2014 alone.
2. Job growth attributable to the rapid development of modern service industries. The service sector contributed more to job growth than either the primary or secondary sector. For instance, there was a great increase in new forms of employment, such as in Beijing's express delivery and "designated driver" (*daijia*)[4] services.
3. Profound shifts in the relationship between supply and demand for labor. China's working-age population has begun to shrink, both in proportional and in absolute terms, making it highly likely that China will experience a structural labor shortage in future.

[4] *Daijia* (literally "substitute driver") has been translated as "designated driver (services)." Basically, an inebriated user would use a car-hailing app (Didi, for instance) to "hail" a "driver on behalf" that will drive them home in their own car. In English-speaking countries, this would probably be known as "drive home services" or "valet services." However, *daijia* is unique to China in terms of both scale and pricing—such services are far cheaper, and hence more popular, in China than in most Western nations.—Trans.

Theory-based estimations indicate that there will be large numbers of unemployed rural residents. However, given the ageing of the rural labor force, it will be difficult for these workers to be effectively deployed in the industrial sector, given that such jobs generally require younger workers. When designing employment policies, China must pay great heed to these new trends. Wage levels will continue to increase, and it will be necessary for China to adapt correspondingly by intensifying workforce training schemes and focusing on labor productivity as a new source of growth. China still has great potential in this respect.

Given that employment is closely related to livelihoods, it is inadvisable for China to lower her guard even though the labor market has improved somewhat. There is still a need to boost employment and promote entrepreneurship, especially when it comes to graduate employment, re-deployment of workers laid off as a result of policies to cut production capacity, and re-employment of workers laid off from small and medium-sized businesses facing economic challenges. The provision of aid to unemployed individuals must continue to be a priority in social governance. In addition, with the changes in the labor market, a younger generation of migrant workers will intensify their demands for greater rights, leading to a higher incidence of labor disputes, which will make it necessary to further develop channels for social mediation and law-based resolution.

Third, shifts in income distribution as China transitions between different stages of development. In general, income gaps have widened steadily in China since her reform and opening up, as the result of both the laws of a market economy and economic policies orientated toward economic efficiency (which had the unfortunate effect of increasing income inequality). This trend peaked in 2008, when the Gini coefficient reached 0.491.

2008 to 2014 saw a gradual decline in the Gini coefficient, which was the result of three factors. First, the urban–rural disparity had begun to shrink. Per capita net rural incomes had grown at a higher rate than urban disposable incomes for four consecutive years. Second, regional disparities have been kept in check. The relatively underdeveloped local economies of central and western China had been growing at a higher rate than the more developed eastern regions for nearly a decade. Third, large-scale poverty alleviation efforts have come to fruition. Income inequality has been worsening among most major nations, including the USA, Russia, and India; only China and Brazil have managed to buck this trend. The improvement in China's income distribution has provided new conditions that are advantageous for her transition toward a model of economic growth and development based on greater consumption. However, income inequality in China remains high relative to other countries. China needs to vigorously address the problem of income inequality, given that it not only holds back economic growth and stymies economic fairness, two of China's development goals, but has also become a deep-seated cause of various social problems.

Fourth, shifts in the employment structure during China's different stages of development. The tertiary sector contributed 50.5% to total GDP in 2015, exceeding 50% for the first time, which was a significant sign that China was about to transition from the "middle stage of industrialization" to the "late stage of industrialization." Because employment flexibility was greater in the tertiary sector than in the secondary sector (which was in turn higher than in agriculture), this shift in the economic structure was also reflected by changes in the employment structure. Alternatively, we may say that the employment structure in China was perhaps on the verge of a new stage, i.e., the "white-collar age," where the majority of the labor force will be employed in white-collar jobs. The white-collar age was also when ideas such as the "age of high mass-consumption" and the "middle class" were first proposed in Western countries. Despite being somewhat controversial, these topics attracted widespread discussion. The formation of a middle class, on the one hand, contributes to social stability and harmony, as well as the formation of mainstream values. On the other hand, it also entails the arrival of an age of diversity and individualism, where new approaches to social governance will be required.

Fifth, shifts in household consumption during China's different stages of development. Since the turn of the twenty-first century, final consumption expenditure and household consumption expenditure have successively declined for more than a decade in proportional terms. However, these trends have been shifting in recent years; in particular, final consumption expenditure and household consumption expenditure have been contributing more strongly toward economic growth. In 2015, final consumption accounted for as much as 66.4% of GDP growth, becoming the most significant driver of economic growth. Although China's age of "imitative, wave-style consumption"[5] has basically ended, diversified, individualized mass-consumption has yet to reach its full potential. In particular, there has been rapid development in new forms of consumption, such as telecommunications, leisure, travel, retirement, housekeeping, health care, education, personal fitness, and online shopping. Consumers have begun to put greater store on the quality of life, especially in aspects such as health, food safety, air and water quality, sense of contentment, and happiness. The age of high mass-consumption has also seen an ever-increasing tendency toward consumerism, leading to economic materialism, "money-mindedness" and widespread conspicuous consumption. Any economic contraction will then have the potential to easily arouse broad discontentment and a relative sense of deprivation. This is a potential problem that social governance in China must remain vigilant against.

Sixth, shifts in population ageing trends during China's different stages of development. As of 2014, 14.9% of the Chinese population, more than 200 million, are above the age of 60. Population ageing has become a global problem in the 21st century. As of today, China has the largest population of elderly persons in the world, more than 250,000 Chinese citizens crossing the threshold into "old age" each day. China's population ageing is characterized not only by sheer numbers but also speed. In addition, young people are increasingly drawn toward urban areas, causing rural areas to experience more severe population ageing than urban areas.

[5] "Imitative, wave-style consumption" (*mofangxing pailangshi xiaofei*) refers to trends in consumption characterized by wide-reaching fads that ebb and flow in popularity. For instance, household appliances were once in great vogue, spawning large numbers of companies that produced largely homogenous products, which inevitably led to market imbalances when something else came into fashion.—Trans.

Population ageing also has made increased pressures on elderly care in China. As the average size of a Chinese family shrinks, the traditional mechanism for elderly care, where seniors live with their children, has been threatened, making it ever more important to build a social retirement safety net. Large-scale institutionalized elderly care remains a rarity—most seniors will continue to spend their golden years living at home. There is an urgent need to develop social services for elderly persons in the community, such as catering, health care, home care, and emergency services. These will require innovative social governance models that involve a greater role for the non-governmental sector, in order to lower the costs of social governance and provide better social services.

The characteristics of China's transition between different stages of development, on the one hand, indicate that Chinese society will continue to undergo momentous social changes even after the economy transitions to the "new normal,"[6] featuring medium-high growth rather than fast growth. In this respect, China still has huge structural flexibility and room for development. On the other hand, it indicates that China will require new strategies, policies, and initiatives to respond to the new, radically different, issues and challenges that have arisen during the course of her development.

A Shift in Focus: From Standard of Living to Quality of Life

When a country or society reaches a certain stage of development, it is inevitable that the people will come to place a greater emphasis on the quality of life. This stronger emphasis on quality is also one of the fundamental attributes of the new growth stage and closely linked to its other characteristics.

Before reform and opening up, China's development emphasized *xian shengchan, hou shenghuo* ("first production, then life")[7] and neglected the role of consumption in boosting the economy. In the context of economic development, this policy orientation manifested as an emphasis on heavy industry over light industry, the latter being closely related to the material needs of daily life. Ultimately, it resulted in a disconnect between production and consumption—there were shortages in nearly all daily necessities. Even basic sustenance was sometimes a problem. After reform and opening up, China reversed direction, opting to base economic development on the production of daily necessities, which resulted in a rapid rise in incomes and consumption. Back when the economic reforms of 1978 had just been initiated, the bicycle, wristwatch, and sewing machine were known colloquially as the *san da jian* ("three big things") which were deemed to be representative of the living standards

[6] The "new normal" generally refers to a stage of growth that emphasizes the quality and sustainability of economic development, in contrast to China's breakneck growth in the past. While the inherent limitations of economic forecasting make it difficult to put a number on "medium-high" growth, it is generally understood as a rate of growth somewhere between 6% and 7%.—Trans.

[7] Basically meaning that raw economic development should take priority over the quality of life.—Trans.

of the people. In the 1990s, the original "three big things" had been supplanted by the television, refrigerator, and washing machine. Today, housing, cars, and insurance have become the new "three big things." The speed of change has exceeded all imagination.

The first hallmark of a greater emphasis on the quality of life is the greater value placed on food safety. When economic reforms were first initiated, the Chinese people had nearly zero awareness of food safety. At the time, basic sustenance, colloquially known as *neng chi bao* ("able to fill your stomach"), was the benchmark for the good life. In fact, the universal standard for measuring rural living standards was "amount of grain per capita." After basic sustenance was no longer a problem, "eating well" (*chi hao*, meaning better nutrition) became the new pursuit. Per capita spending on non-staple foods (as a proportion of total food expenditure) became an important indicator for whether people were "eating well." After better nutrition became the norm, people came to place greater store on other forms of consumption, such as education, travel, telecommunications, leisure, and so on. At this time, "food expenditures as a proportion of total spending," i.e., Engel's coefficient, an internationally adopted benchmark, became the basic indicator for measuring living standards. Now, at a stage of development characterized by the pursuit of quality of life, people demand not only to "fill their stomachs" and "eat well," but also to "eat organic" and "eat natural." A series of food safety scandals have caused people to view food safety as an important factor in ensuring a greater quality of life.

The second hallmark of a greater emphasis on the quality of life is the greater value placed upon the environment, particularly air quality. During the childhood years of most Chinese alive today, azure skies dotted by white clouds and starry nights were the norm. Sadly, most people failed to appreciate these—the environment did not seem to have much to do with the quality of life. After reform and opening up, the rapid pace of economic growth caused an immense increase in environmental pollution. Still, most people accepted that this was a necessary price for higher incomes and greater living standards. "Pollute first, clean up later" also seemed to be a path of development that was hard to sidestep.

Yet, the course of history can sometimes be altered by a single incident. In mid-January 2013, the city of Beijing endured successive days of severe smog, which led authorities to issue an "orange alert" (the second-highest level), which caused large numbers of air travelers to be stranded at airports. Much of central and eastern China, from the northeastern and northern regions to the Huang-Huai[8] and Jiangnan regions, was blanketed in air pollutants. In some regions, visibility was less than one hundred meters. Among the 120 key cities monitored by the Ministry of Ecology and Environment, 67 were deemed to be polluted, while 22 expressways across eleven provinces were partially closed. This incident led to nationwide reflection. What are we pursuing? What is the point of development? What was the point of pursuing living standards or quality of life if the air was so terrible that one had to don masks when outdoors? A new social consensus was taking shape: "lucid waters and lush mountains are invaluable assets."

[8] Referring to the region south of the Yellow River and north of the Huai River.—Trans.

The third hallmark of a greater emphasis on the quality of life is the greater value placed upon health and personal fitness. As living standards improved, average life expectancy became an important indicator for measuring the quality of life. The increase in life expectancies was not only due to better living standards, but also due to improvements in the healthcare safety net and advances in medicine. The large, fatal pandemics of the past, including leprosy, bubonic plague, smallpox, schistosomiasis, dengue fever, and various viral influenzas, have gradually been conquered by the human race. However, new fatal diseases such as cancer have reared their ugly heads. Health has become an important aspect of the quality of life, which has given a great boost to health-related industries. Poor health has become an important factor that adversely impacts the quality of life.

The fourth hallmark of a greater emphasis on the quality of life is the greater value placed upon greater social engagement. The transition from an agrarian society to an urban society is also a transformation from a "society of acquaintances" to a "society of strangers." A society's level of psychological care is often linked to the loneliness and depression suffered by people living in a society of strangers, as well as the psychological stress caused by fast-paced modern life. Although psychological care is not well-developed in China, interpersonal relationships, which are important in Chinese culture, often act as a substitute. Nonetheless, against the backdrop of rapid social development, increasing numbers of people are suffering from loneliness, depression, and paranoia, the result of living among a society of strangers. At the same time, people are also developing an ever-stronger awareness of liberty, rights, and participation, which have made social expression, social trust, social support, social justice, and social engagement important conditions for ensuring a greater quality of life.

The fifth hallmark of a greater emphasis on the quality of life is the greater value placed upon emotional well-being. For a long time, quality of life was assessed solely on the basis of material living conditions and associated welfare indicators. Today, people have come to place an unprecedented emphasis on happiness and contentment.

A Review of Previous Research on the Quality of Life

The American economist John Kenneth Galbraith was one of the first researchers to explore the quality of life from a theoretical perspective. In *The Affluent Society*, published in 1958, Galbraith argued that quality of life should not only measured in terms of private wealth. Quality of life also referred to the enjoyment or pleasure, that people derived from creature comforts, "conveniences of life" and "spiritual fulfillment." Of course, Galbraith also pointed out that, although happiness was a universal pursuit, there remained great uncertainty with regard to actually *measuring* happiness. He stressed that, at the time, there was a stark contrast between private wealth and the filthy public environment in America, arguing that the latter was a major impediment to quality of life. Galbraith characterized it thus:

> The family which takes its mauve and cerise, air-conditioned, power-steered and power-braked automobile out for a tour passes through cities that are badly paved, made hideous by litter, blighted buildings, billboards, and posts for wires that should long since have been put underground. They pass on into a countryside that has been rendered largely invisible by commercial art. (The goods which the latter advertise have an absolute priority in our value system. Such aesthetic considerations as a view of the countryside accordingly come second. On such matters we are consistent.) They picnic on exquisitely packaged food from a portable icebox by a polluted stream and go on to spend the night at a park which is a menace to public health and morals. lust before dozing off on an air mattress, beneath a nylon tent, amid the stench of decaying refuse, they may reflect vaguely on the curious unevenness of their blessings.

Galbraith believed that, after attaining personal prosperity, the daily demands of the population would pivot toward a focus on the quality of public services—this made it necessary to correct the erroneous notion that wealth was solely generated by the private sector. After its publication, *The Affluent Society* was immensely popular for a time. The USA had just declared a "war on poverty," and *The Affluent Society* became recommended reading for scores of college students.

Galbraith wrote *The Affluent Society* against a specific historical backdrop. The renowned contemporary American historian Eric Foner dedicated a chapter titled "An Affluent Society" (1953–1960) in *Give Me Liberty: An American History*, with sections respectively titled *A Changing Economy, A Suburban Nation, The Growth of the West, A Consumer Culture, The TV World, A New Ford, Women at Work and at Home, A Segregated Landscape, Public Housing and Urban Renewal*, and *The Divided Society*. Foner described the "Golden Age" experienced by Americans then in the following passage:

> On the surface, the decade of the 1950s seemed uneventful. It was a time of widespread affluence in the United States, the beginning of an unprecedented economic expansion that lasted until the early 1970s. Millions of Americans moved to the suburbs, where they enjoyed accessed to an astonishing array of consumer goods that poured out of American factories, including cars, television sets, and household appliances…
>
> In every measurable way—diet, housing, income, education, education, recreation—most Americans lived better than their parents and grandparents had. By 1960, an estimated 60% of Americans enjoyed what the government defined as a middle-class standard of living. The official poverty rate, 30% of all families in 1950, had declined to 22% a decade later (still to be sure, representing more than one in five Americans).

It was also then that some American academics and government officials launched the "social indicators movement" in an attempt to incorporate comprehensive social indicators into the formulation of development strategies, evaluation of national conditions, social planning, social policies, and assessment of the quality of life. The overarching goal of the movement was to move beyond the traditional practice of measuring development and living standards solely using economic indicators.

In the 1950s and 1960s, American academia saw great enthusiasm for research into quality of life. In 1957, Gerald Gurin, Joseph Veroff, and Sheila Feld of the University of Michigan, in collaboration with other research institutions, conducted a nationwide randomized survey which focused on studying the mental health and happiness of Americans. In 1965, Hadly Cantril published a comparative study that measured levels of general satisfaction across 13 countries (including the US). Almost at the same time, Norman Bradburn conducted a nationwide survey in America, where respondents were asked to rate their levels of happiness.

Social Indicators, a collection of academic papers edited by Bauer in 1964, focused on the indirect impact of national spatial planning on American society. These research findings aroused widespread interest in the quality of life. Rostow closely examined related issues in his 1971 book *Politics and the Stages of Growth*, which defined the "age of high mass-consumption" as a higher stage of development that would succeed the take-off stage.

Back then, the University of Chicago's National Opinion Research Center and the University of Michigan's Institute for Social Research were the two main university-based polling centers in the USA. These institutions were active participants in large-scale questionnaire surveys on the quality of life and played a key role in collecting relevant data. Thereafter, the quality of life gradually became a dedicated field of research pursued all over the world. In fact, the field was deeply influenced by the practice of measuring social quality using a system of comprehensive social indicators.

In the 1990s, research institutions all over the world began to review the practice of simply using individual economic indicators to measure standards of living and development. The United Nations Development Programme (UNDP) proposed the Human Development Index (HDI) in the Human Development Report 1990, which measures the socioeconomic development of the various UN member states. The HDI forms a stark contrast with the World Bank tradition of ranking countries by development standards using gross national product (GNP). The HDI is a comprehensive index based on three variables: life expectancy at birth, the "education index," and "a decent standard of living."

Since 1990, the UNDP has published the HDI for each country annually and in so doing played an important role in guiding the corresponding development strategies adopted by developing countries. In 1997, the European Commission pointed out that it was inadequate to solely rely on economic stimulus when addressing the issues that have surfaced in the various EU member nations, as a focus on economic policy would not resolve structural problems faced by the EU. The political, economic, and social stability of the various EU member nations could be effectively safeguarded only by revisiting social policies.

It was in this context that the EU issued the *Amsterdam Declaration on the Social Quality of Europe*, which pointed out that the various EU member nations should work toward increasing their employment rates, alleviate poverty, and increase the healthcare and social resources available to the public. The declaration also specified that improving "social quality" should be a main target of EU policies. In addition, the declaration also clearly states that social quality was to be defined by indicators in four respects: social security, social cohesion, social inclusion, social empowerment. The theoretical framework for social quality differs from the theoretical framework for quality of life or the HDI. Social quality focuses more on the interactive process between individual development and social development. In social quality theory, overall social development is quantitatively assessed by measuring economic and social indicators, while the quality of social development is assessed from the perspective of social safety nets, the individual's level of social inclusion, their "life chances" (or opportunities for personal development) and their level of individual agency.

After reform and opening up, with the rapid rise in living standards, Chinese scholars also began to pay attention to the quality of life. As a result, the importance of social indicators came to be increasingly appreciated. In 1983, China's National Bureau of Statistics proposed a draft outline concerning social indicators and began to publish relevant time series data from 1984 onwards. In 1985, Professor Lin Nan, a Chinese American sociologist, and the Institute of Sociology at Tianjin Academy of Social Sciences jointly conducted a questionnaire survey that covered 1,080 households in Tianjin. In 1987, Professor Lin teamed up with the Institute of Sociology at Shanghai Academy of Social Sciences to conduct yet another survey on a similar scale. Later, based on these data, the team would develop structural models that studied the relationship between quality of life and social indicators. In October 1987, at a seminar on social reforms and "lifestyle theory" convened in Tianjin, several academics would explore in greater detail the concept of quality of life, related indicators, and the contemporary state of relevant research in foreign academia. Thereafter, quality of life became a common topic of research and discussion among Chinese sociologists.

Beginning in 1988, the Institute of Sociology at the Chinese Academy of Social Sciences, during a research project on social development and social indicators, organized together with partner institutions, proposed five groups of indicators for measuring the overall social development in a particular region. Later on, a research project led by Zhu Fangqing, a researcher at the Institute, would track and evaluate the quality of life and level of social development in China for several years, ultimately proposing a system of indicators that included household consumption, income, fulfillment of basic needs (food, clothing, and so on), energy consumption, "conveniences of life" (*shenghuo fangbian chengdu*), and "spiritual fulfillment" (*jingshen shenghuo*). In 1989, a research group from the Institute of Sociology at the Jiangsu Academy of Social Sciences studying modernization and "new socialists" distributed questionnaires in urban and rural areas of five Chinese provinces (Jiangsu, Henan, Jilin, Sichuan, Guangdong), asking residents to rate their quality of life.

From 1987 to 1990, several sample surveys were conducted in parts of the municipalities of Beijing, Xi'an, and Yangzhou by a research group from Peking University's Department of Sociology that was pursuing research into the quality of life. Apart from including objective indicators, the researchers also added "reference indicators"[9] to measure other factors that impacted an individual's perception of his

[9] "Reference indicators," in this context, refers to the measures against which an individual assesses his quality of life. For instance, when asking respondents to rate their present living standards, the researchers also asked them to rate the living standards that they enjoyed five years ago (i.e. the "reference indicator"). They found that an individual's subjective rating of their present circumstances was correlated with such "reference indicators." In a similar vein, they also discovered that "keeping up with the Joneses" had a measurable effect on how satisfied an individual is with their quality of life.—Trans.

quality of life. In addition, the researchers "consolidated" various objective indicators using "intermediary evaluative indicators."[10] Ultimately, they obtained a three-level[11] model that explained the role of subjective and objective factors in influencing the quality of life.

In December 1991, a group of researchers engaged in comparative research on the population and quality of life in China, which included members of Fudan University's Institute of Population Research, Nanjing University's Population Research Institute, and the Institute of Population Economics at the former Beijing Institute of Economics,[12] launched an academic seminar on the quality of life, attended by scholars from all over China. The attendees discussed definitions of "quality of life," relevant indicators, as well as ways to measure quality of life, eventually coming up with many useful insights. Based on the conference papers, the first monograph on quality-of-life research was published in 1992.

Research related to the development of a "moderately prosperous society" has pushed research on the quality of life in China to a new level. In 1991, researchers from the Institute of Sociology at the Chinese Academy of Social Sciences, including myself, Lu Xueyi, and Zhu Qingfang, published a book titled "China's Moderately Prosperous Society in 2000," wherein we proposed a system of social indicators for China's *xiaokang* society which included sixty indicators spanning six major categories and compared these with the standards used in other countries. In 2003, Zhu Qingfang and I published another book, titled *Build a Well-Off Society in China*,[13] which further elaborated upon an indicator system and standards for evaluating China's progress toward a "moderately prosperous society in all respects." The book also comparatively evaluated the various provinces and regions of China, as well as fifty-seven major municipalities. During this period, the National Bureau of Statistics also officially released a set of sixteen indicators, known as the *xiaokang shehui zhibiao tixi* [Indicator System of a Moderately Prosperous Society], as well as a more comprehensive set of twenty-three indicators known as the "Statistical Indicator System of Building a Moderately Prosperous Society in All Aspects."

The greater emphasis on quality of life has promoted research in this area, whereas such research has in return motivated reflections upon the standards used to measure quality of life. By reviewing past academic literature on the quality of life, we may perhaps draw the following conclusions. Firstly, "quality of life" is a concept distinct from standards of living. Generally, a better quality of life is built upon higher living standards, but the former is not wholly based upon the latter. Secondly, among the

---

[10] Basically, respondents were asked subjective questions that covered an entire range of objective indicators. For instance, objective indicators such as the size of a respondent's dwelling, the number of people in the same household, and so on, were "consolidated" into one subjective survey question: "Do you feel that you live in a crammed dwelling?"—Trans.

[11] The three levels are respectively, in ascending order, "objective indicators," "subjective ratings," and "satisfaction with a particular aspect of life" (such as quality of housing).—Trans.

[12] In 1995, the Beijing Institute of Economics merged with other institutions to become the Capital University of Economics and Business.—Trans.

[13] "Well-off" here also refers to *xiaokang*—at the time, translations were not as standardized.—Trans.

material conditions that determine the quality of life, an increase in living standards also cause changes what people demand and expect in life. Certain qualities that are easily neglected when living standards are low, such as air quality, insolation, traffic conditions, daily conveniences, population intensity, and crime rates, have become important factors that impact the quality of life. Thirdly, as living standards increase, subjective indicators, such as happiness, sense of satisfaction, and social identity, have also become indicators that are key to measuring the quality of life. At the same time, these subjective indicators remain connected to particular social structures, social relations, and social systems.

Analytical Approach and Content Framework

The emergence of social indicator analysis has greatly promoted research into quality of life. Assessing the quality of life through social indicators makes it possible to compare the different qualities of life in various countries, regions, and cities. This allows us to rank different localities based on their residents' quality of life, which often attracts considerable media coverage that leads to widespread attention. However, this enthusiasm for using social indicators to formulate rankings has the tendency to lead researchers in less-than-ideal directions, where they prefer to expend their energies on descriptive indicators rather than inquiries into the causal factors that impact the quality of life.

All assessment systems for social indicators suffer from a fatal weakness, i.e., that it is near impossible for them to be entirely independent. Regardless of what methods are deployed when selecting indicators, and regardless of how many (or few) indicators are ultimately selected, the results cannot be said to be entirely "scientific" or "exhaustive." Given that different scholars hold different theoretical assumptions, the indicators that they select will be different, which inevitably impact the outcomes of their rankings.

Research on the quality of life should return to analyzing and uncovering the underlying laws governing a society's development, which requires that researchers focus their efforts on various specific fields of research. Certainly, compared to rocket scientists who can accurately plot a missile trajectory, social scientists cannot be expected to have as acute a grasp of the laws behind the intricate workings of society, because there are too many complex, uncontrollable factors that underlie social phenomena. Of course, social phenomena are not entirely random and arbitrary. There are deep-seated laws that govern the trends by which they develop.

Our research is based on the analysis of data acquired through large-scale sample surveys. However, we do not aim to establish a comprehensive system for evaluating social quality. Neither do we hope to evaluate the quality of life in different regions or cities. There has been much previous research on these topics—at present, it is difficult to imagine how such approaches may lead to new research breakthroughs. In contrast, we hope that this book may demonstrate our understanding of the concept of "quality of life" in terms of both breadth and depth and establish a theoretical framework for related analysis, in order to explain the disparities between reality and what is predicted by social indicators. For instance, why do people shy away from localities that have been deemed to have a higher "quality of life"? Why are reported "levels of happiness" higher in some localities deemed to have a lower "quality of life"?

Our analytical approach in this book covers four types of factors that impact the quality of life: (1) objective factors, such as income, consumption, employment, education, housing, and health care; (2) subjective factors, such as happiness, sense of satisfaction, perceptions of environmental quality, and judgment on whether society is "good"; (3) social factors, such as social justice, social trust, social support; (4) social mechanisms, such as social mobility, social participation, and social cohesion (see Fig. 1). The chapters of this book are largely organized in accordance with this framework.

Analysis of Objective Factors:

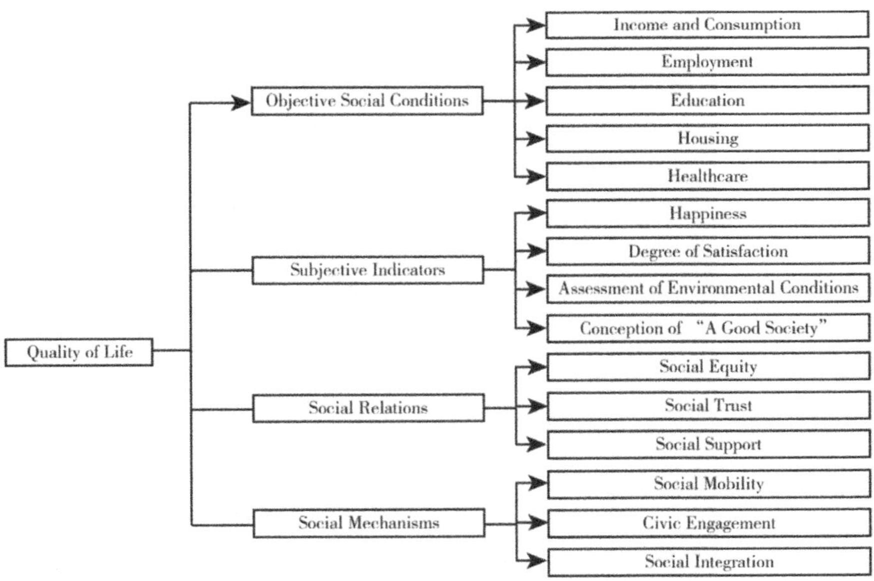

Fig. 1 Quality of life—an analytical framework

Peilin Li

Contents

Chapter 1
Consumer Stratification and Quality of Life

Yi Zhang

In order to prevent China from falling into the "middle-income trap" amidst downward pressures on the economy, policy makers were forced to seek a balance between three imperatives—exports, investment and consumption—and design relevant economic policies in a targeted manner to ensure their consistency with current economic development goals. However, given that a weak global economy has led to sluggish exports and investment, domestic demand has become the main force driving the overriding imperative of economic growth. Meanwhile, the Chinese consumer market is transitioning from "imitative, wave-style consumption"[1] towards multi-tiered, individualized, and diversified consumption patterns, as President Xi Jinping pointed out during his address at China's Central Economic Work Conference in December 2014. This situation, on the one hand, reflects the progressive upgrading of consumption in China after the country entered the "middle-income stage". On the other hand, it also implies that social stratification will lead to segmentation in consumer markets.

Thus, national masterplans designed to stimulate the economy need to be closely in tune with the consumption preferences of different social groups, while also taking into account the need to improve the quality of life across the board given the uneven income distribution in China. On this basis, I analyze the marginal propensity to consume of different social groups from the perspective of social stratification and provide proposals for policies that stimulate consumption. In addition, specific means for improving the quality of life will be discussed, with reference to the consumption characteristics of different social groups.

Meanwhile, when producers and distributors target different groups with products of different price ranges, qualities, and features, consumers will gradually be divided

Y. Zhang (✉)
National Institute of Social Development, Chinese Academy of Social Sciences, Beijing, China

[1] See note in Introduction—Trans.

P. Li (ed.), *The Quality of Life in Contemporary China*, Research Series on the Chinese Dream and China's Development Path, https://doi.org/10.1007/978-981-16-7021-3_1

into "hierarchical categories" based on their purchasing preferences. This is a fundamental conclusion derived from sociological research. Both "Conspicuous consumption" (or the consumer demand from the "leisure class"[2]) and non-rational forms of consumption that have been "infused" with cultural symbolism[3] are consumption trends that are an inevitable outcome of social stratification. Given these circumstances, the formulation policies for stimulating consumption and domestic demand is greatly served, in terms of both theory and practice, by treating social groups as a variable in research on social stratification in China, the different consumption propensities of different social groups, and their quality of life.

1.1 Literature Review, Data, Research Methodology and Main Variables

(1) Literature Review

In his 1971 book, *Politics and the Stages of Growth*, Rostow sought to explain the demand of the masses for a higher quality of life by proposing that "the age of mass-consumption" would be succeeded by "the search for quality". According to the tenets of Rostow's theory, China should be at a turning point between what he called "the drive to maturity" and "the age of high mass-consumption". When discussing the peculiarities of consumption in Chinese society, Wang Ning also pointed out that Chinese cities are gradually becoming "consumer societies", even though China remains, on the whole, a "dual-track" society where mass production and "elite consumption" co-exist (Ning 2009).

It can be said that the dawn of the industrial society liberated human beings from the material scarcity of agricultural societies. Large-scale mechanized production gave rise to new forms of supply and a wider range of consumer goods. Thus, industrial societies and post-industrial societies gradually became consumer societies. For instance, Simmel argued that, given income constraints, when the upper classes emphasize the "heterogeneity" of their consumption, the middle class and the "underclass" would bring themselves closer to the upper class by means of imitating trends in "fashion".[4] This is also as Baudrillard argued: in a consumer society, consumption of material goods has transformed into a symbol of social status. Hence, in the

[2] "Potlach" is a concept described by the American anthropologist Franz Boas. His student Ruth Benedict clarified that the concept was more closely related to "Kwakiutl potlatches", serving as an analogy for extravagantly wasteful consumption that serves to show off one's wealth and set one apart from others. In *The Theory of the Leisure Class*, Thorstein Veblen variously described consumption by the leisure class as "conspicuous consumption", "visible consumption", and "vicarious consumption".

[3] In *The Cultural Contradictions of Capitalism*, Daniel Bell analyzes "rational" and "non-rational" forms of consumption when discussing Veblen's research.

[4] See *Philosophie der Mode* [The Philosophy of Fashion] by Georg Simmel.

"grand narrative", consumption would become status symbols,[5] where people begin to emphasize the symbolic value of consumption goods over their use-value. This phenomenon was also discussed incisively by Daniel Bell in *The Cultural Contradictions of Capitalism* (Bell 2010). Although later sociological research on consumption, particularly the postmodernist literature, came to emphasize various other aspects, whether the influence of the mass media or the autonomous choice of the consumer, they were largely in agreement as regards to how consumption in industrial societies serve as symbols of identity and status.[6]

Description and analysis of a consumer society from a qualitative perspective profoundly characterizes the various transitions from an agricultural society to an industrial society to a post-industrial society, from a society of material scarcity to a society of material abundance, and from a closed society to a globalized society, as well as the outcomes thereof. However, the fluctuations inherent in a market economy lead to extremely complex outcomes.

When the economy is booming, the supply of consumption goods automatically generates demand, in a fashion similar to that described by Say's law. However, during economic downturns, a slowdown in consumption will cause existing surpluses to be exacerbated. Thus, given the segmented nature of consumer markets, more quantitative research is needed to inform the formulation of optimal masterplans for satisfying or stimulating consumer demand, and enact institutional arrangements that can be accepted by each social group, especially those less affluent. After all, levels of happiness depend on the satisfaction derived from either consumption itself or relative equity in consumption levels.

The German statistician Engel was one of the first to use quantitative indicators to describe how consumption patterns vary across social groups. In so doing, Engel discovered that food expenditures decreased as incomes increased. This is precisely why the UN Food and Agriculture Organization has used Engel's coefficient to divide consumers into different tiers, i.e. "extremely poor", "barely subsisting", "moderately well-off", "affluent", and "extremely affluent".[7] Since the initiation of economic reforms, China's Engel's coefficient has been progressively declining, from 57.5% in urban areas and 67.7% in rural areas in 1978 to 39.2 and 49.1% respectively in 2000, and then again to 35.0 and 37.3%.[8] When using Engel's coefficient as a means for identifying different social groups, researchers have also taken into account the impact of the Asian Financial Crisis on the Chinese market. In a paper titled "Consumption Stratification: An Important Perspective for Stimulating Economic

[5] See *The Consumer Society* by Jean Baudrillard.

[6] Pierre Bourdieu, 1984, *Distinction: A Social Critique of the Judgement of Taste*, London: Routledge & Kegan Paul. In research on class distinction and the "symbolism of consumption" in post-industrial societies, some post-modernist scholars believe that consumption is no longer significant to class distinction. However, this does not undermine the classical sociological theory that consumption plays a role in class distinction in industrial societies.

[7] The UN FAO classifies social groups by their Engel's coefficient into "extremely poor" (above 0.60), "barely subsisting" (0.50–0.59), "moderately well-off" (0.40–0.49), "affluent" (0.30–0.39), and "extremely affluent" (below 0.30).

[8] As determined by data from the *China Statistical Yearbook* for the relevant years.

Development", Li Peilin and Zhang Yi discussed how social status, as an objective variable, can inform policies for stimulating domestic consumption (Li and Zhang 2000).

But there are certain limitations to Engel's coefficient. First, in the long run, Engel's coefficient will trend downwards as incomes grow. In the short run, however, Engel's coefficient may trend either downwards or upwards.

Second, a society's Engel's coefficient is generally stable. However, under the impact of certain factors, profound changes in the supply side may lead to shifts in consumption patterns. An increase or decrease in other forms of consumption may result in strong fluctuations in Engel's coefficient and cause it to lose its ability to explain differences in the quality of life. For instance, a family that takes out a mortgage on a home may see a strong decline in its Engel's coefficient due to the increase in mortgage payments. By the same principle, when rural residents build new homes, they lower their spending on food.

In these circumstances, a decrease in Engel's coefficient does not necessarily indicate a rise in current quality of life. In fact, it may even reflect temporary declines that result from lower levels of spending on food or other forms of consumption.[9] In reality, many people have a tendency for skimping on basic necessities to purchase expensive "durable consumer goods". Furthermore, in many scenarios, mortgages take a long time to pay off even as the broader economy continues to fluctuate. During economic downturns, if mortgage payments cannot be reduced even as current or expected income levels decline, a lower Engel's coefficient would become entirely irrelevant to measuring quality of life.

A rise in Engel's coefficient can be caused by a rise in food prices, a decline in non-food prices, or a broadening income gap caused by greater levels of growth in the income of affluent social groups. This means that there are times when rising incomes across the board are not necessarily accompanied by absolute declines in Engel's coefficient.

Hence, based on the fundamental principles behind Engel's coefficient, it is more meaningful to divide consumption into the consumption of basic necessities and consumption that serves future development, which would allow us to apply Engel's coefficient to analysis of consumption trends and purchasing behavior when discussing changes in household consumption patterns. In reality, when studying income structures and consumption patterns, many scholars separately consider income derived from different sources and consumption through different channels to allow for more detailed analysis.[10] This is because a more detailed breakdown of income and consumption allows us to more clearly discern trends of change in

---

[9] Apart from housing, there are many other factors that may lead to a rise or decline in Engel's coefficient. These include expenses on education, healthcare, or elderly care for the infirm, all of which lower proportionate spending on food.

[10] Research by Li Shi and Luo Chuliang discovered that different levels of expenditure on education and healthcare not only have an impact on current income gaps, but also on future income gaps, given that such expenditures have a deep impact on children's future opportunities. See Li Shi and Luo Chuliang. "Changes in Household Income Inequality in China: Short-term Variation and Long-term Trend", *Comparative Economic and Social Systems*, no. 4 (2012).

each form of consumption (as a proportion of total consumption). This also facilitates better research on the impact of different forms of expenditure on consumers' current and future quality of life.

We observe a clear hierarchy in consumption behavior because of the hierarchical nature of human needs. In his 1891 foreword to *Wage Labour and Capital*, Engels categorized the means of consumption into "the means of life, of the enjoyment of life, and of the development and activity of all bodily and mental faculties". In his theory on the hierarchy of needs, Maslow also classified human needs into the categories of physiology, safety, belonging and love, esteem, and self-actualization. In Maslow's conception, humans can only pursue needs of a higher level after satisfying needs of a lower level. Therefore, it is consumption that satisfies basic needs that is most significant to less affluent social groups—the pursuit of future development is possible only if life itself can be sustained.

Hence, it would be more helpful to policy makers if we were to quantitatively break down household consumption into survival-oriented consumption (consumption necessary for subsistence to sustain production and reproduction[11] through work) and developmental consumption[12] (consumption that improves the quality of life and one's future development). This not only helps to stimulate consumer demand, but can also ensure greater social equity in consumption patterns.

In contrast to economists, who analyze consumption patterns using an abstract unit (i.e. money), sociologists focus on preference and location when studying the consumption of specific goods by different social groups (Wang 2007; Zhu 2012). For instance, it is common for surveys to ask respondents where they dine (choices include expensive restaurants, food courts, ordinary establishments, and street food), where they purchase their clothes, whether their household owns a computer, television, refrigerator, washing machine, automobile, and so on.[13] The basic frameworks of sociological research assume that those who dine in expensive restaurants and purchase branded clothing are of higher social status and enjoy a better quality of life. This undoubtedly serves to identify class distinctions in a buyer's market, or even during the transition from a buyer's market to a seller's market.

But when there is a supply glut, or when a seller's market emerges, because of the smaller differences between the use-value of various consumer goods, their value as status symbols will exceed their use-value. Hence, it would be ineffective to simplistically base class distinctions on the types of consumer goods. For instance, there are great gaps in the price and quality of cars, televisions, refrigerators and computers. Class distinctions that are only approximately based on such categorical variables would be highly unpersuasive. However, it would be undoubtedly far too cumbersome to ask the respondent to specify the brand, price, product specifications, energy consumption, and location of purchase of every durable good owned by

[11] Referring to the Marxist concept of "economic reproduction"—Trans.

[12] Consumption for the enjoyment of life may be categorized under developmental consumption.

[13] Many recent surveys have included questions regarding expenditures on dining, apparel, and household appliances. See the various CASS and CAS surveys.

their household. In addition, such data is extremely hard to obtain.[14] Surveys of consumption on branded goods and antiques show that, in the context of markets saturated with counterfeit products, it becomes even more difficult to sieve the chaff from the wheat and make a better judgment on the quality of data.

In such circumstances, if we ignore the different tastes of various social groups, assuming instead that branded consumption goods with cultural symbolic value command greater prices, and that prices can reflect differences in spending power, we may study consumption preferences and patterns by identifying different social groups through household spending power, or the amount of money that a household spends on certain types of products or services during a certain period. This would also facilitate our discussion of various market-oriented stimulus policies. In addition, greater social equity should be a fundamental focus of such policies.[15] Gary Becker, through quantitative analysis of relevant models, vividly described consumption by a household or an individual as "a process of production or reproduction" whereby income utility is maximized from the satisfaction derived from spending on consumer goods that serve as status symbols (Becker 1987, pp. 9–10)

(2) Data

This article uses data from the Chinese General Social Survey 2013 by the Institute of Sociology at the Chinese Academy of Social Sciences. The CGSS 2013 collected data from 10,206 respondents that were randomly selected, including details on their socioeconomic status, income and consumption, and household expenditures on food, clothing, utilities, rent, financial support for parents, weddings, funerals, education, travel, healthcare, entertainment, household appliances, mortgage payments, phone bills, and commuting. This allowed researchers to analyze respondents' consumption patterns.

(3) Research Methodology

In research on the sociology of consumption, some scholars focus on the actual consumption of goods to analyze how consumption levels, preferences, and trends differ across social groups. Some researchers also determine levels of real consumption for each social group through levels of nominal consumption, which then allows them to analyze the impact of consumption on quality of life.

In this chapter, I determine levels of actual consumption by each social group based on nominal consumption expenditures. We then demonstrate the relationship between income and consumption through an analysis of their average propensities to consume. Thereafter, we demonstrate how income growth can lead to changes in demand. This allows us to make policy proposals informed by class analysis.

---

[14] Bourdieu used "taste" as a variable in research on the consumption preferences of different social groups to divide respondents into "avant-garde" and "traditional".

[15] Any expenditure on stimulus policies should fundamentally serve to enhance social equity. Hence, any government investment in this respect should have safeguards against exacerbating the social divide between affluent and less affluent social groups.

(4) Main Variables

First, from the perspective of operationalization, we assume a respondent's socioeconomic status reflects that of their household. Some researchers have argued that this may not always be true, but have been unable to propose feasible alternatives. Meanwhile, much research has shown that a father's socioeconomic status is positively correlated with the socioeconomic status of his children, while a husband's socioeconomic status is highly correlated with that of his wife. Hence, I divide respondents into the class of business owners, new middle class, old middle class, working class, and farmers based on their "employment status" (i.e. whether they are part of the class of business owners), level of professional skill required by their job, and their "power" in the workplace.

The class of business owners mainly refers to those who do not engage in labor but instead employ others to work for their companies. This class can be further subdivided into "owners of large businesses" (those who employ more than ten people) and "small businesses owners" (those who employ seven or fewer).[16] However, we do not make this distinction in our chapter as China does not have a large class of business owners.

The middle class can be divided into the new middle class and the old middle class. In class analysis, some researchers make this distinction while others often do not. But as a matter of academic principle, it is necessary to divide the middle class into the new middle class (referring to those employed in white collar jobs) and the old middle class (referring to persons in various forms of self-employment), because of the great differences in their individual backgrounds, level of educational attainment, and working environment. The main reason for making this distinction in the present chapter is that I believe there are differences in their level of consumption.

The working class are those blue-collar workers that are employed in various industries. This includes both semi-skilled workers and menial laborers.

Farmers are defined as members of households primarily engaged in agriculture, forestry, animal husbandry, and fishery, i.e. those who derive a living from their own land through the household contract responsibility system. Here, it is important to note that a respondent who is a worker employed in these industries is categorized as part of the working class. Respondents who own agricultural businesses that employ agricultural workers are categorized as part of the class of business owners.

Second, we define household expenditures per capita on food, clothing, utilities, rent, healthcare, maintenance of parents, weddings, and funerals to be consumption that serves basic household needs (i.e. survival-oriented consumption), while household expenditures per capita on education, travel, entertainment, household appliances, phone bills, and commuting are defined as "developmental consumption". In general, a household would consider how much should be spent on developmental consumption only after taking care of survival-oriented consumption.

---

[16] In his research, Erik Wright, a prominent analytical Marxist sociologist, once divided members of the "capitalist class" into those who hired fewer than 10 employees and those who hired 10 or more.

When treating consumption as a variable, we consider consumption expenditures from the perspectives of average and marginal propensity to consume. The average propensity to consume is the ratio of a household's current consumption to current income, while marginal propensity to consume refers to how current consumption expenditures are affected by changes in income.[17] I first use an initial model to analyze the overall average and marginal propensity to consume for each social group before using an expanded model to analyze their marginal propensities for survival-oriented and developmental consumption.

1.2 Average and Marginal Propensities to Consume for Each Social Group

In all societies, income distribution has an impact on consumption. In addition, income is also redistributed at the household level. Family altruism implies that family members with higher incomes would transfer their incomes to family members with lower or no incomes (Becker 1987). Hence, the per capita consumption of a household plays an important role in the study of consumption. Though the purchase of consumer goods may be based on individual preferences, total consumption in any given period is necessarily constrained by total household income.[18] In this sense, class analysis accounts for constraints on spending power.

After dividing members of society into the class of business owners, new middle class, old middle class, working class, and farmers, we discover that the class of business owners, new middle class, old middle class, working class and rural class respectively make up 4.67, 12.98, 13.92, 33.60 and 34.83% of the total labor force. Thus, we see that 30-odd years of economic reforms have led to major transformations in China's social structure. If we define the "middle class" to include the class of business owners, new middle class, and old middle class, we find that this "middle class", the working class, and farmers, would each makeup roughly one-third of the total labor force. Thus, we see that a predominantly rural society has become one evenly dominated by the middle class, working class, and farmers. This change in social class structure is fundamental to understanding the current policy landscape in China.

The transition from the planned economy to the market economy also brought vast and rapid changes to class structure in China. These changes have also reinforced overall trends in social transformation, reflecting a strong trend towards a "worker-based" and "middle-class-dominated" society. It can be said that the size of the working class in China is now the largest that it has ever been as a percentage of the total population, while the middle classes (both old and new) are growing rapidly.

[17] Per Keynes' absolute income hypothesis, $C_t = a + bY_t + U_t$, where C_t represents household consumption in period t, a represents autonomous consumption (the vertical intercept, in graphical terms), Y_t represents income for the current period, and b represents marginal propensity to consume.

[18] For our purposes, a person living alone is considered as a "one-person household".

In other words, growth in China's working and middle classes have coincided in the same time and space. In future, with advances in agricultural modernization, increased rates of land circulation, and increased educational enrollment, the size of the farming population will shrink in both absolute and relative terms.

In addition, as both megacities and large cities assume the characteristics of post-industrialized societies, the absolute size of the working class will stabilize after reaching a certain level. The working class will also cease to grow as a proportion of the total labor force once there is no need for further industrialization.

The old middle class will also stabilize after reaching a certain size, just as in other East Asian countries strongly influenced by Confucian culture. Only the new middle class will continue to grow in size to make up an ever-greater proportion of the total labor force. As in any other market economy, it is impossible for China's class of business owners to make up a large proportion of the labor force.

This means that growth in the new middle class will lead to future transformation in China's consumer markets. However, this transformation will be progressive in nature. As the Chinese economy develops, investment will contribute less to economic growth. Instead, as China deepens its ties with global markets, and with continued tensions in global trade, fluctuations that result from growth driven by external demand will also become more prominent. In such circumstances, domestic demand will come to play a greater role in economic growth than in any other period.

However, widening income gaps will lead to greater divides in Chinese society. Although figures from the National Bureau of Statistics show that the Gini coefficient

Table 1.1 Population structure and consumption propensity among different social groups (initial model)

Social group	Size as proportion of labor force (%)	Average propensity to consume (household) (%)	Marginal propensity to consume (household)
Class of business owners	4.67	44.47	0.177*
New Middle Class	12.98	76.23	0.471*
Old Middle Class	13.92	83.24	0.513*
Working Class	33.60	83.94	0.395*
Agricultural Workers	34.83	101.53	0.315*

Notes
(1) Actual size of the household is not considered, given that we assume that household socioeconomic status is equivalent to the socioeconomic status of the "head of the household" (i.e. the *huzhu*, literally "master of the household". The *huzhu* is generally the senior male member of the household—Trans.)
(2) $*p < 0.001$

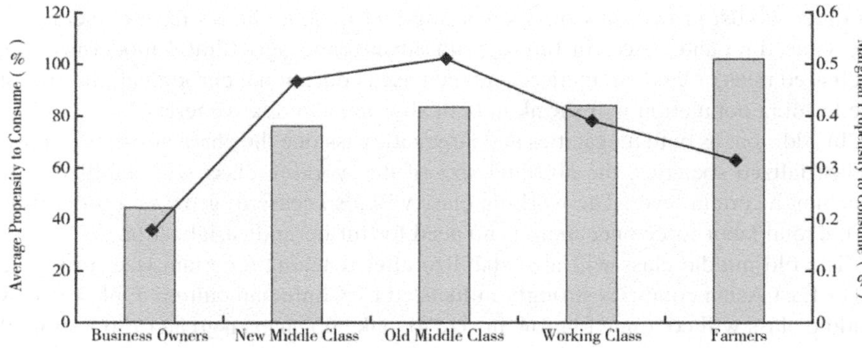

Fig. 1.1 Average and marginal propensities to consume by social group

has been on the decline, it was still as high as 0.469 as of the end of 2014,[19] suggesting that the income gap remains rather large in China. This is also demonstrated by differences in the average propensity to consume, which is determined by income levels. As illustrated in Table 1.1, the average propensity to consume of the class of business owners, new middle class, old middle class, working class, and farmers is respectively 44.47, 76.23, 83.24, 83.94 and 101.53%. Farmers have an average propensity to consume greater than 100% because some poor households have to spend on basic necessities even if they have no income in the current period. This is in line with Keynes' "law" that average propensity to consume declines as incomes grow.

This means that those with higher incomes—the class of business owners—tend to spend less of their current incomes on consumption, instead saving or investing the surplus. In contrast, social groups with lower incomes must direct their current income towards basic needs to maintain standards of living that reflect the levels of social development.

However, it was not necessarily true that households of higher socioeconomic status had a lower marginal propensity to consume. Instead, marginal propensity to consume was lower for farmers and the class of business owners, but higher for the social groups in between (see Fig. 1.1). This means that income growth leads to greater growth in consumption for some social groups compared to others. From the CGSS 2013, we observe that farmers and the class of business owners had a lower marginal propensity to consume than the new and old middle classes and the working class.

Trends revealed by the data have significant policy implications.

First, to stimulate economic growth through domestic demand, it is necessary to first increase the income levels of those less affluent. Although those from more affluent households would consume more, on a per capita basis, than their less affluent

[19] *Zhonghua Renmin Gonghe Guo 2014 Nian Guomin Jingji he Shehui Fazhan Tongji Gongbao* [PRC 2014 National Economic and Social Development Statistics], http://news.xinhuanet.com/ttgg/2015-02/26/c_1114446937.htm.

counterparts—consumers of higher socioeconomic status would consume more in absolute terms—such consumers are small in numbers and have a lower average propensity to consume. This means that their levels of consumption will cease to grow after reaching a certain level.

This also implies that, when there is a large income gap (i.e. a higher proportion of income goes towards the owning and middle classes), consumption by less affluent social groups will be limited, while affluent social groups are less motivated to consume as their needs have been fully met.

When domestic consumer goods can no longer satisfy the demands of affluent social groups, they will develop a taste for foreign travel and shopping, where international markets can provide them the "shopping satisfaction" that domestic markets cannot. In so doing, these consumers "convert" domestic demand into foreign demand, which has a strong stimulus effect on global markets but a very weak stimulus effect on domestic markets. To encourage domestic spending by affluent social groups, it is necessary to enhance the competitiveness of domestic brands and their substitutability with foreign products. If it takes too long to enhance the value and competitiveness of domestic household products or daily consumer brands, it would become impossible in the near future to reverse the preference of affluent social groups for foreign shopping, which would in turn have an adverse impact on policies that seek to stimulate economic growth through domestic demand.

In addition, data on the average propensity to consume indicate that farmers, the new and old middle classes, and the working class all have strong demand for consumption. Potential for growth in consumption is greatest for farmers and the working class. As their combined strength exceeds 2/3 of the total labor force, an increase in their income levels will not only enhance overall spending power but also lead to a greater rise in the overall quality of life.

In the long run, domestic demand would be more greatly stimulated by growth in the incomes of less affluent social groups. However, efforts to increase the incomes of farmers and the working class would take a long time to bear fruit. It is difficult for a society to adjust income structures once income distributions have taken form and mechanisms that impact income distribution have ossified. Furthermore, given that enterprises' overall management and technical capabilities remain low, policies that increase their labor costs would also impact their competitiveness in the international arena. This means that changes to consumption structure rely on changes to income distribution, while changes to income distribution rely on changes to the industrial structure, which in turn rely on the rate of technological innovation.

Second, the tendency of marginal propensities to consume to rise to a certain point before falling (i.e. the so-called inverted-U curve) not only suggests that income growth among farmers and the class of business owners does not lead to an increase in their consumption levels. Greater incomes may lead to greater expenditures on some goods and services but not others. Income elasticity of consumption is greater for the working class and the new and old middle classes, with the old middle class exhibiting the strongest marginal propensity to consume.

One major factor is that the old middle class is mainly comprised of older rural migrants who have acquired shopfronts and become self-employed. In a sense, they have become "urbanized" over the years, but are still officially labelled as rural migrants. After having become accustomed to urban life, most are unwilling to return to the villages. Though it is difficult for them to fully integrate into urban life, they have set down roots in cities with their unique "grassroots characteristics", and are hence more inclined towards spending on property and consumer goods.

Meanwhile, why is it that farmers have a strong average propensity to consume, but a relatively weak marginal propensity to consume? One possible explanation is that some of the goods produced by farmers are reserved for self-consumption, for instance vegetables, grains, and fruits. In addition, the inadequacies of the social safety net, in particular medical insurance and old-age insurance, may compel farmers to set aside their own "insurance". At present, old-age insurance schemes for rural residents are vastly inferior to that for urban workers. Although rural medical insurance has improved since the introduction of the New Rural Cooperative Medical Scheme, reimbursement rates remain low. Furthermore, a sizeable proportion of the medicines dispensed in many participating hospitals are not covered under insurance.

To a large extent, these factors constrain the spending power of farmers. In addition, rural households are generally frugal, while in keeping with the Chinese tradition, elderly parents in rural areas are dependent on their children for care. However, today's population is far more mobile, and their adult children have found themselves drawn to the cities.

The new middle class, being more highly educated, has higher levels of human capital, as well as relatively stable incomes and social security. Their greater spending power results in a stronger marginal propensity to consume. The middle class has largely moved on from the simple pursuit of use-value. Amidst the wave of globalization, they have become more focused on the symbolic value of consumption. We can predict that the new middle class will become even more inclined towards symbolic value as China becomes a post-industrial society. As the use-value of consumer goods take a backseat to their symbolic value, both Chinese and foreign manufacturers would have to adopt targeted strategies in their competition for customers in a stratified market.

Given that the new middle class is growing rapidly, the old middle class will stay largely at the same size, the working class will grow until a certain scale, while the population of farmers will shrink, China will have to focus more on the targeted marketing of consumer goods to different social groups in an era of higher income levels. China will also have to develop its own consumer groups amidst competition with global manufacturers and brands. It would be difficult to China to enhance the competitiveness of its products if manufacturers cannot refine their target markets.

1.3 Marginal Propensities for Survival-Oriented and Developmental Consumption

A rise in the quality of life is none other than a transition from survival-oriented consumption to developmental consumption. This means that if people spend most of their income on survival-oriented consumption, it would be impossible for them to enhance their quality of life. If people can progressively break through the constraints of survival-oriented consumption during the transformation of consumption patterns, allowing developmental consumption to increase in both absolute and proportional terms, they would gradually come to enjoy a greater quality of life. Hence, in this sense, the quality of life is evaluated both objectively and subjectively (by oneself or others).

This is also precisely why a person may have a very high level of (subjective) life satisfaction, but a relatively lower quality of life, as measured objectively. In reverse, a person's objective quality of life may be high, but his or her subjective level of life satisfaction may be lower. Why do people in certain regions report greater levels of happiness despite their somewhat low standards of living? One important factor is that their expectations of life are close to their actual living standards. A greater gap between high expectations and low living standards lead to greater negative utility. Quality of life can only be objectively compared in this sense.

In our former discussion, survival-oriented consumption is defined as per capita household consumption on food, clothing, utilities, rental, healthcare, maintenance of parents, weddings, and funerals, while developmental consumption is defined as per capita household consumption on education, travel, entertainment, home appliances, phone bills and transport. When survival-oriented consumption makes up a greater proportion of spending, people are said to be in a state of survival-oriented consumption. When developmental consumption makes up a greater proportion, people are said to be in a state of developmental consumption. Obviously, quality of life is more readily enhanced for social groups with a greater marginal propensity for developmental consumption.

From Table 1.2, we observe that, in the expanded model for survival-oriented consumption, farmers, the working class and old middle class show a stronger marginal propensity to consume (respectively 0.249, 0.241 and 0.302), while marginal propensity for survival-oriented consumption by the new middle class is only 0.168. The class of business owners has a particularly weak marginal propensity for survival-oriented consumption (0.002; statistically insignificant). It can be observed that in current Chinese society, consumption demand from farmers and the working class is dominated by survival-oriented consumption. Constrained by their lower income levels, these groups spend more in proportionate terms on clothing, food, housing, as well as utilities, healthcare, maintenance of parents, weddings, and funerals.

Here, we must note that it is unthinkable for a Chinese villager to avoid contributing towards the wedding and funeral expenses of friends and family. In a "society of acquaintances", certain ceremonial rituals are key to maintaining

Table 1.2 Marginal propensity for survival-oriented and developmental consumption (expanded model)

Survival-oriented consumption					
	Farmers	Working class	Old middle class	New middle class	Class of business owners
(Constants)	3144.867	3092.476	4892.054	3233.185	7155.003
Household income per person	0.249	0.241	0.302	0.168	0.03
Household residency (urban = 1)	215.516	1280.11	−459.356	1638.880	2221.005
Household savings per person (including creditor's rights)	−0.016	0.041	−0.030	0.020	0.078
Average years of schooling per household member	246.553	995.813	64.349	357.643	230.585
Average current value of durable consumer goods and collectibles	0.039	0.056	0.017	0.000	0.061
Average current value of operating assets per household member	0.001	0.018	0.000	−0.005	−0.002
N	1830	1562	688	584	195
Adjusted R^2	0.130	0.292	0.248	0.273	0.552
Developmental Consumption					
	Farmers	Working Class	Old Middle Class	New Middle Class	Class of Business Owners
(Constants)	370.617	−955.624	−765.375	−4255.860	−7965.284
Household income per person	0.096	0.146	0.204	0.257	0.002

(continued)

Table 1.2 (continued)

Survival-oriented consumption					
	Farmers	Working class	Old middle class	New middle class	Class of business owners
Household residency (urban = 1)	97.827	430.309	−609.307	4847.203	2891.371
Household savings per person (including creditor's rights)	−0.033	−0.012	0.017	−0.029	0.073
Average years of schooling per household member	364.917	873.236	822.760	1564.047	4921.152
Average current value of durable consumer goods and collectibles	0.168	0.236	0.144	0.054	0.278
Average current value of operating assets per household member	0.043	0.047	0.000	0.005	−0.003
N	1872	1612	710	598	200
Adjusted R^2	0.92	0.236	0.225	0.098	0.231

Note In the expanded absolute income hypothesis model, we have controlled for household residency, household savings per person, average years of schooling per household member, average current value of durable consumer goods and collectibles, and average current value of operating assets per household member

interpersonal relations, friendships, and family networks. A villager can skimp on other current expenses but not spending required by *renqing*.[20] In China's social networks, structured tightly by a "differential mode of association", people contribute a "respectable" sum of cash to such events. Here, what is considered "respectable" is determined by their family ties (by blood or marriage) or friendships. In addition, "face" is an important consideration. These factors make it near impossible for

[20] *Renqing*, literally "personal feelings", can be understood as a Chinese social norm where individuals are expected to fulfil the obligation of reciprocity, i.e. consideration for the feelings of others—Trans.

those living in rural areas to cut back on such expenditures.[21] So-called reciprocity in *renqing* also lead to greater wedding and funeral expenditures.[22] Furthermore, as farmland increases in value and the "development of a new countryside"[23] picks up pace, rural households are also investing more in housing. The transition of rural dwellings from mud/wood structures to brick, and then to reinforced concrete is also a sign of overall social progress.

Poor families whose yards and houses "have not kept up with the times" would face pressure to "keep up with the Joneses". Funding from local governments in recent years for rural roadworks and renovation of decrepit buildings have also provided incentives for the improvement of rural infrastructure and the construction of new housing. Care for elderly parents, including medical expenses, are also a major part of rural household spending. Although the New Rural Cooperative Medical Scheme has already guaranteed, to a very large extent, access to healthcare resources for rural residents, hospitals that command their confidence are mostly located in county seats and regional urban centers. Given these circumstances, an elderly person who is hospitalized must be accompanied by several other persons who take turns to provide care, which results in expenditures other than medical bills. For instance, a caregiver would need to stay in a hotel, purchase food, pay for non-reimbursable medical tests, and so on. To a certain extent, this increases the financial burdens on farmers and limit their demand for developmental consumption. Hence, farmers have a relatively low marginal propensity for developmental consumption (only 0.096).

The working class is currently seeing a rise in their quality of life. While the working class is predominated by rural migrants, SOE workers continue to make up a sizeable proportion.[24] SOE workers enjoy stable incomes and enjoy protection from "wage adjustment policies", where wages are adjusted based on changes in local price levels. In recent years, pensions for urban workers have increased by 10% annually, which has given eligible enterprise workers greater confidence in a financially secure retirement. This has allowed the working class to gradually become more confident consumers, which has raised their quality of life. However, there are only so many SOE workers. Although industrial workers who were once rural migrants have seen their nominal incomes grow despite market fluctuations, higher costs of living in urban areas (food and rent) have caused them to become less confident consumers.

[21] Many localities have been unsuccessful in their efforts to encourage less elaborate weddings and funerals. One important reason is that traditional family and social networks are still a strong source of social support. Even in modernized urban communities, people cannot avoid such expenses if they live among a "community of acquaintances", given the need to gain the approval of their peers and maintain their social networks.

[22] The size of a gift can only spiral upward, given that one is obligated to never give less than one last received—Trans.

[23] Referring to the promotion of village modernization in China, referred to in policy documents as "building a new socialist countryside" (*jianshe shehuizhuyi xin nongcun*)—Trans.

[24] Enterprise reforms in China over the late 1990s and early 2000s led to a decline in the number of "collective enterprises" (enterprises owned by a collective). SOEs have declined in number even in prefecture-level cities. Most workers in collective enterprises have either found new employment in private enterprises or have retired after becoming eligible for a "seniority buyout" (*maiduan gongling*)—Trans.

Meanwhile, a young male rural migrant seeking marriage finds it hard to find a spouse if he does not [at least] own a home in his hometown's county seat.[25] The "mother-in-law" factor in rising housing prices means that a prospective son-in-law (or his parents) has to exhaust their household savings to purchase a home, which then lowers their spending power. A rural migrant, or even his entire family, needs to scrimp and save for many years just to foot a down payment on an apartment. Future mortgage payments would then suppress their future spending. Because certain levels of survival-oriented consumption must be maintained, it will be difficult for these groups to spend more on consumption that raises their quality of life. It must also be emphasized that consumption by members of the working class who are rural migrants is limited by their inability to fully integrate into their city of residence. Being uncertain of their future prospects, these workers move from city to city without thought of permanent settlement. As such, they have no choice but to focus on survival-oriented consumption instead of developmental consumption. For migrant workers, separation from their family members also inhibits their households' motivation to consume. This is reflected in our models, where the working class has a marginal propensity to consume of merely 0.146, higher than farmers but lower than the new and old middle classes.

The old middle class, i.e. those who earn a living through self-employment, are mostly elderly rural migrants, while some are technicians who have recently started their own businesses. Those members of the working class who were once migrant workers must rely on their own menial labor and a certain level of technical skills to sustain their living standards and employment. In this sense, an employee's physical stamina and skills—i.e. their human capital—depreciate over time. Therefore, in a discriminatory market environment, it is increasingly difficult for older migrant workers to find jobs in manufacturing or other similar fields. Increasingly, they are "forced" into unskilled construction jobs, and eventually "forced" to return to the countryside. Some migrant workers have even dyed their hair in fear of being seen as too old to keep their jobs in the cities.[26]

Even so, migrant workers have been forced to seek other careers as they advance in age. Those unwilling to return to the countryside as farmers choose to start businesses in urban areas of their hometowns or cities where they used to work, where population ageing provides them with greater opportunities. Some farmers whose land has been requisitioned have joined the "low-end" service industry after receiving their compensation monies. Because of their position at the "low end" of the service industry and intense competition from their peers in the same social group, their incomes are relatively limited. However, as they are no longer separated from their families, households of the old middle class have the strongest marginal propensity

---

[25] In China, nearly all parents demand that a prospective son-in-law (or his parents) purchase a matrimonial home for the couple as a condition for agreeing to the marriage. This demand is often more strongly pressed by the mother, which has led to the common saying in China that "housing prices are driven by mothers-in-law" (*zhangmuniang tuidong fangjia*)—Trans.

[26] *Chengshi Luntan* [City Forum]: *Woguo Gaoling Nongmingong huo chao 4000 Wan Kao Ranfa Duo Jiancha* [More than 40 million migrant workers in China dye their hair to evade checks], http://www.chengshiluntan.com/forum.php?archiveid=2&mod=viewthread&tid=5338318.

for survival-oriented consumption, as well as an inclination towards developmental consumption. For instance, their marginal propensity for developmental consumption is only 0.204, second only to the new middle class.

Meanwhile, living standards are rapidly rising among the new middle class. In general, members of the new middle class are white-collar workers that have at least a college graduation diploma and earn more than farmers, the working class, and the old middle class. Financial pressures are greater for fresh graduates that have just joined the new middle class, who need to purchase homes in the city where they work. However, their spending will no longer be constrained by survival-oriented consumption after they have purchased a home. On the whole, they have a relatively low marginal propensity for survival-oriented consumption but a relatively strong marginal propensity for developmental consumption (0.257). Hence, trends of consumption upgrading and greater "personalization" in consumer goods are prominently manifested among the new middle class. They are also well aware of the value of human capital, making them willing to invest in an overseas education for their children. They value the symbolic value of consumer goods and are willing to spend on branded goods. During the "household appliances revolution", they frequently replace their home appliances. They are also the main spenders on automobiles, not only turning China into a nation "on wheels", but also stimulating the tourism sector through greater demand for road trips. As major spenders on urban hotels and dining, they have also given a boost to "weekend consumer markets". These factors explain why their marginal propensity for developmental consumption is greatest among all social groups (0.257).

However, it is surprising that the class of business owners shows neither a significant average nor marginal propensity to consume. Why would there be no statistically significant changes in the consumption of a theoretically wealthy social group, especially in terms of marginal propensity to consume, after controlling for such variables as operating fixed assets per household member? My initial hypothesis was that members of the class of business owners who hired employees had, on average, incomes that were higher than other social groups, implying that they would spend more on developmental consumption, which would be reflected in our model as a greater marginal propensity for developmental consumption. However, this has been refuted by our data. Why? One possible explanation is that they label spending on developmental consumption as business expenses, a common practice among business owners. For instance, money spent on certain home appliances, personal vehicles, or a new apartment[27] (registered to the company), may be listed as fixed assets of the company, while expenses incurred by the business owner or their family on overseas travel, education, entertainment, phone bills, and commuting may also be disguised as ordinary business expenses. Furthermore, it is impossible for some small family businesses to separate household expenditures from business expenses.

[27] Even in large cities such as Beijing, Shanghai, Guangzhou, or Shenzhen, there are no official government restrictions on the number of apartments that a company may purchase.

Hence, statistical insignificance in the data does not necessarily suggest that this social group has a lower marginal propensity for developmental consumption.

Our analysis shows that changes in China's social structure have created a stratified consumer market. Marginal propensity for survival-oriented consumption is greater among farmers and the working class, while the old middle class has a higher marginal propensity for both survival-oriented and developmental consumption. Meanwhile, the new middle class has become more focused on developmental consumption. The *nouveau riche* class of business owners may have already moved on from developmental consumption to "hedonic consumption". Although there are frequent media reports regarding conspicuous consumption by the class of business owners, the CGSS 2013, regrettably, did not include questions that could have collected information relevant to any further analysis.

Levels of marginal propensity for survival-oriented and developmental consumption by each social group also have great policy significance. On one hand, this information allows governments to formulate more targeted policies for stimulating consumption. On the other hand, such data allows producers to more accurately target consumer demands for products and services that enhance their quality of life.

Firstly, farmers and the working class are the main forces driving survival-oriented consumption, while the old middle class drives both survival-oriented and developmental consumption. With greater economic and social development, these classes have been able to greatly improve their living standards through greater consumption. By improving the quality of life, economic reforms have also raised average life expectancy in China to approximately 75 years.[28] However, socioeconomic progress has also raised poverty thresholds and average living standards to new heights. In contrast to yesteryear, full stomachs, warm clothing, along with televisions and gas supply in every home, have become the norm in rural communities.

Greater freedom of land circulation and the building of apartment complexes in rural villages will also allow farmers will come to enjoy greater levels of developmental consumption. The rollout of the "electrical appliances to the village"[29] and "automobiles to the village"[30] policies in 2008 will also encourage greater levels of

[28] "Zhongguo Renkou Pingjun Yuqi Shouming Da 74.83 Sui, Shi Nian Tigao 3.43 Sui" [Average life expectancy in China reaches 74.83 years, increasing by 3.43 years over the past decade], http://news.sohu.com/20120810/n350317599.shtml.

[29] In December 2007, pilot trials of a policy initiative known as *jiadian xiaxiang* ("electrical appliances to the villages") were launched in the provinces of Shandong, Henan, and Sichuan. Under this policy, rural households were entitled to a 13% subsidy when purchasing eligible home appliances. Nationwide implementation followed in February 2009. Later on, the Ministry of Finance, Ministry of Commerce, and Ministry of Industry and Information Technology announced in a circular that the policy would come to an end on January 31, 2013.

[30] According to the *Automobile Industry Adjustment and Revitalization Plan* (*Qiche Hangye Tiaozheng Zhenxing Guihua*) issued by the State Council on January 14, 2009, from March 1 to December 31 of 2009, one-time subsidies will be provided for the purchase of small passenger vehicles (1300 cc or below) and the replacement of three-wheelers or low-speed cargo vehicles with light cargo vehicles. In early 2010, the policy was extended until December 31 of the same year.

developmental consumption by rural households. After completion of the "consumption revolution" for the working class, where it was encouraged to replace obsolete home appliances with new models, home renovation may become the focus of a new round of consumption stimulus.

As a matter of fact, changes in China's social structure will also lead to major structural shifts in food consumption. In the past, most spending on food went towards staples, i.e. grain, whereas meat, poultry, eggs, dairy, and vegetables are forming a larger part of food consumption. The conventional practice of relying on greater use of chemical fertilizers and pesticides to increase food production is now being replaced by a trend towards eco-friendly farming practices. Healthcare is also seeing progressive improvement.

The urban middle class and class of business owners have already enjoyed the fruits of such transitions, while rural households and the working class have come to be more aware of food safety and nutrition than during any other period in history. However, rural households, the working class, and the old middle class continue to spend less on dairy and meat products than the new middle class and owing class. At the same time, as the population continues to age, healthcare and nursing for the elderly will also become key sectors for developmental consumption, the size of which will be astounding even without additional stimulus. Currently, health product advertisements have become a staple in mass media broadcasts, but these products are rarely as effective as advertised. There will be great future demand for such products.

Secondly, the class of business owners, along with the old and new middle classes, will become the main spenders on developmental consumption. These social groups drive the upgrading of urban consumer goods, regardless of the size of the cities they live in. As these social groups continue to increase their development consumption, they have come to demand more customized goods of greater quality, allowing foreign products to become widely popular in China. Foreign enterprises have come to dominate all sorts of markets, from luxury cars to phones and stylish wallets. This suggests that the main contradiction faced by Chinese enterprises is the shortage of adequate developmental consumer goods in the face of huge consumer markets and greater consumption demand by the class of business owners. Take the sales of mobile phones in 2015 as an example. Despite a slowing economy, pre-orders for Apple's iPhone 6S and iPhone 6S Plus exhausted stocks within merely a few days of their launch. As advertisements and marketing discourse pushed "use value" to second place behind "status value", competition in the consumer markets have greatly intensified. Furthermore, markets for developmental consumption, including health, wellness, entertainment, and travel, have not been adequately developed. Even in localities where these industries are relatively mature, accessibility remains limited due to inadequate transportation networks. In addition, it is imperative for policy makers to consider means for raising developmental consumption by enhancing work efficiency, shortening working hours, and increasing the "supply of leisure" (i.e. more holidays and free time during weekends).

1.4 Conclusion and Brief Discussion

Based on figures released by the National Bureau of Statistics in September 2015, retail sales of consumer goods in August increased by 10.8% over the same period in 2014,[31] which boosted expectations that consumption rather than investment will become the main driver of economic growth.[32] However, a closer analysis makes it apparent that consumption has become a greater driver of economic growth only because investment and exports have shrunk, whereas sustained development requires maintaining a certain level of domestic consumption. Hence, greater stimulus is required if economic growth is to be driven mainly by domestic consumption. The upgrading of consumption is both a result of socioeconomic development and necessary for a higher quality of life across the board. As China continues to progressively enhance its consumption structure, an analysis of the marginal propensity to consume by each social group and their spending on survival-oriented and developmental consumption leads to the following basic conclusions.

First, for social groups that are less well-off, income growth is key to greater consumption and a higher quality of life. We observe from their marginal propensities to consume that consumption potential is greatest for the rural households, the working class, and the old and new middle classes. Demand for improving current quality of life through consumption is strongest among these social groups. Yet, their spending power is limited by stagnant income growth. Without rapid income growth, consumption growth comes to rely on the expected level of future "stability" or "insurance". However, the "insurance" function of the social safety net remains inadequate. As the economic landscape becomes ever more complex, any simplistic, "sloganistic" proposal for increasing the incomes of less well-off social groups will be largely impractical over the short run.

It is no longer effective for rural households to increase their incomes simply through food production, given that international food prices are generally lower than domestic food prices and the limited amount of arable land per capita. Direct food production subsidies, grants of agricultural supplies, and "quality variety subsidies"[33] remain limited. The policy effects of the New Rural Cooperative Medical Scheme and old-age insurance, which are funded by transfer payments, have largely been exhausted. While policies for raising old-age pensions for urban workers by 10% each year have been implemented for a decade, monthly old-age pensions for rural residents have merely increased from CNY 55 to CNY 70 in 2015. Meanwhile, inflation is chipping away at the level of security provided by such forms of "wide-coverage, low-benefits insurance".

---

[31] Guangming Wang [gmw.cn]. "8 Yue Woguo Shehui Xiaofeipin Lingshou'e Tongbi Zengzhang 10.8%" [Consumer retail sales in China increased by 10.8% in August over the same period last year], http://news.gmw.cn/newspaper/2015-09/14/content_109182517.htm.

[32] In 2014, consumption accounted for 50.2% of China's GDP growth. In the first half of 2015, consumption contributed 60% towards GDP growth.

[33] Known as *liangzhong butie* in Chinese. Farmers are eligible for subsidies if they cultivate certain crop varieties encouraged by the state—Trans.

As a slowdown in economic growth makes it difficult for the central government to sustain fiscal subsidies over the long run, growth in rural incomes must rely on agricultural modernization. The ageing of the rural population is also undermining the sustainable development of the agricultural industry and rural areas, as well as sustainable growth in living standards for rural households. Growth in rural incomes will rely on higher labor productivity, as well as policies that allow for effective land circulation, which would make more arable land available to individual farmers by reducing the size of the rural population.

The working class has seen significant income growth over the past decade. The average migrant worker now makes more than the average fresh graduate. It has also become the norm for a skilled laborer to earn far more than a university graduate. By "voting with their feet" and "resistance of the disadvantaged", the labor-capital relations prevalent in the late twentieth century no longer apply to today's migrant workers, who have greatly improved their working conditions. However, China's working class still finds it hard to compete with their international counterparts, given that companies are slow in upgrading their technologies, while levels of skills and automation remain low, which limits the possibility of achieving income growth through mechanisms for "tripartite negotiation".

Hence, sustained income growth for the working class is preconditioned upon labor productivity growth rates. Stubbornly high urban housing prices have led to high mortgage payments and rent, leading the working class to lament that they are "working for the banks". Inflation is also a far greater constraint upon consumption by the less well-off classes than their wealthier counterparts. Given such circumstances, if the costs of survival-oriented consumption remain high, the working class would find it impossible to increase spending on developmental consumption, making it difficult to achieve a higher quality of life. As a shrinking population eats away at China's demographic advantages, the working class will have to rely on greater labor productivity to sustain growth in income and consumption.

The old middle class shows demand for both survival-oriented and greater developmental consumption. However, members of this group are employed mainly in the service industries, which narrows any potential growth in their incomes. Often, the physical stores that they run have been rendered obsolete by e-commerce. Those who wish to upgrade their businesses and provide "high-end services" find it difficult to obtain bank loans, while intensified competition in "low-end services" has eroded their profits. Local policies to encourage entrepreneurship have also not lived up to expectations. Despite the hopes of the old middle class to incorporate their businesses and "grow bigger and stronger", the realities of market competition and economic fluctuations have limited their room for development. Neither are they able to generate outsized gains from small operations, as was the case during initial market reforms. Given these circumstances, any growth in income would have to rely on their business acumen, as well as their ability to provide better products and services.

Second, survival-oriented consumption is no longer an issue for the more well-off social groups (i.e. the new middle class and the class of business owners), which have begun to pursue greater developmental consumption. As the first to benefit greater labor productivity brought about by technological advancement, these groups enjoy stable incomes and a greater capacity for risk. They are more concerned with the status value of consumption, even with regards to basic necessities such as food, clothing and housing. Their developmental consumption will focus on brand value. While satisfying their material needs, they will also seek spiritual fulfilment, which would make them more discerning consumers of cultural products and services. China lacks high-quality, expressive literary works, artwork, films, television programs, and stage plays that are steeped in its national history, in much the same way as the country lacks high-quality, personalized, consumer goods with strong symbolic value and "national characteristics".

Thus, we find that Chinese society faces the following conundrum. On one hand we find that the middle class has a low opinion of poorly made, "mass-produced" films and television programs, leading to their low ratings and poor box office turnouts. On the other hand, we find that foreign films generate astronomical box office returns. Given this supply and demand structure, stimulating consumption by the new middle class and class of business owners will require China to address the mismatch between their tastes and the quality of Chinese products. Furthermore, China also needs greater innovation and development in philosophy and the social sciences, in order to better inform the worldview of its people during the course of modernization and post-modernization. When material prosperity is not matched by enrichment of the mind, affluent social groups may come to adopt a solely materialistic view of consumption, giving full rein to their desires at the expense of moral ethics, which would erase any gains from social progress and development.

Third, when it comes to the production and supply of consumer goods, it is important to draw a distinction between survival-oriented and developmental consumption. Consumer goods targeted at farmers, the working class, and the old middle class can remain homogenous and mass-produced, given that their consumer demand is dominated by survival-oriented needs. After all, survival-oriented consumption remains strongly "imitative and wave-style", given that low-income groups are mainly concerned with the use-value of their purchases.

However, the new middle class and the class of business owners have already come to demand consumer goods that more closely fit their individual tastes, which have led to consumption patterns that are more diverse, multi-tiered, and differ strongly between different age groups. This would require the integration of information technology and production technology, i.e. combining the production of consumer goods with the internet. Manufacturers targeting the new middle class should progressively transition from mass production of consumer goods to "batch customization", and then to individualized customization. Consumers are already able to retrieve information on consumer products through the internet, social media, and smartphones, which lays a strong foundation for popularizing such forms of consumption.

The service sector should also seek to meet the preferences of the class of business owners and the new middle class by developing new forms of consumer-oriented products and services, including in the healthcare, education, travel, elderly care, leisure, and environmental industries. As shown in Table 1.3, there are statistically significant differences in the views of different social groups on "whether economic

Table 1.3 How strongly do you feel that "economic development is more important than environmental protection"? (% of respondents)

Social group	Very strongly	Strongly	Not very strongly	Not at all	Unable to say	Total
Class of business owners	5.6	19.1	51.8	20.9	2.6	100
New middle class	5.2	14.0	58.1	21.2	1.5	100
Old middle class	8.1	26.2	47.7	13.0	5.0	100
Working class	7.8	22.9	50.9	14.4	3.9	100
Farmers	8.6	30.7	41.5	8.3	10.9	100
Sub-total	7.6	24.8	48.2	13.3	6.1	100

development is more important than environmental protection". The majority of the class of business owners and the new class favor environmental protection policies, while farmers, the working class, and the old middle class favor economic development. In this sense, social stratification has left a deep mark on markets for consumer goods and services (the environment can be seen as a public good).

Needless to say, quality of life cannot be merely measured by consumption expenditures. Factors such as leisure, work intensity, family harmony, quality of friendships, and levels of comfort should also be included. However, consumption should be the foundation for enhancing the quality of life. In the context of a market economy, a high quality of life cannot be achieved without a certain level of spending on consumption.

References

Becker, Gary. 1987. *A Treatise on the Family*. Translated by Peng Songjian. Beijing: Huaxia Publishing House.

Bell, Daniel. 2010. *The Cultural Contradictions of Capitalism*. Translated by Zhao Yifan et al. Beijing: People's Publishing House.

Di, Zhu. 2012. The methodology and research strategies of mixed methods design and its application: In a case study of consumption patterns [in Chinese]. *Sociological Studies* 4.

Ning, Wang. 2009. *From the Ascetic Society to the Customer Society [in Chinese]*. Beijing: Social Sciences Academic Press.

Peilin, Li, and Zhang Yi. 2000. *Consumption Stratification: An Important Perspective for Stimulating Economic Development [in Chinese]. Social Sciences in China* 1.

Wang, Jianping. 2007. *Zhongguo chengshi zhongjian jieceng xiaofei xingwei [The consumption behavior of the urban middle class in China]*. Beijing: Encyclopedia of China Publishing House.

Xiaolu, Wang. 2007. Huise shouru yu jumin shouru chaju [Gray Income and Residents' Income Gap]. *Comparative Studies* 31.

Chapter 2
Quality of Employment and the Underlying Factors

Lei Fan

As the working age population in China continues to shrink, imbalances in the supply and demand for labor have been somewhat mitigated, leading to lower employment pressures. On one hand, the rapid development of the service sector has increased the number of available jobs. On the other hand, there are still severe structural problems in the labor market. The mismatch between supply and demand will make it difficult to resolve graduate unemployment in the short run, even as there continues to be a significant shortage of highly skilled workers.

Concurrently, the slowdown in economic growth will lead to an even greater shortage of workers in certain regions and industries. Supply will continue to outstrip demand in certain sectors of the labor market. Hence, for the foreseeable future, employment rates will remain a fundamental concern for the government when setting economic growth targets.

While employment rates are still important, employment quality has also been receiving more attention. While measures to encourage informal employment, implemented during the 1990s to ease employment pressures, have effectively boosted employment rates, these same measures have suppressed income levels and led to lower job security. Meanwhile, there are large numbers of migrant workers in urban areas who face poor working conditions and low levels of social security. Difficulties faced by university graduates seeking employment have also directly affected their quality of employment—many graduates have found themselves in want of good quality jobs. Hence, given that China's labor imbalances have been abated, improvements in income levels, working conditions, and social security for workers will not only enhance the overall quality of employment but also further ease employment pressures.

L. Fan (✉)
Institute of Sociology, Chinese Academy of Social Sciences, Beijing, China

P. Li (ed.), *The Quality of Life in Contemporary China*, Research Series on the Chinese Dream and China's Development Path, https://doi.org/10.1007/978-981-16-7021-3_2

2.1 Literature Review

The interaction between workers and the factors of production is reflected in their quality of employment, which includes the nature of work, qualification requirements, salary, job security, working environment, social security, and labor relations (Liu and Han 2007). As economic globalization picks up pace, labor relations are becoming increasingly globalized. The involvement of multinational capital has not only caused labor conflicts in various countries, but has also made it more complex for a country to effectively balance labor-capital relations and protect workers' interests. Meanwhile, global economic downturns have also caused greater unemployment rates and deteriorating quality of employment to become a global problem (Liu and Han 2007). To address such issues, the 87th International Labour Conference that took place in June 1999 proposed the notion of "Decent Work", i.e. promoting decent, productive employment opportunities for men and women, against the backdrop of globalization, that are in line with the conditions of liberty, fairness, safety and human dignity.

Thereafter, international studies on employment quality have made progress in developing systems for appraising job quality at the individual level, comprehensive frameworks for assessing employment quality at the national level, and indicators for assessing job quality at the industry level. This led to the development of several indicator systems, notable examples of which include the "Decent Work" indicators proposed by the International Labour Organization, the "Quality of Employment" indicator framework developed by the United Nations Economic Commission for Europe, and the system of indicators for "quality of job and employment" put forth by the European Science Foundation (Wang 2014; Tian and Man 2013).

Inspired by foreign research on the quality of employment, Chinese scholars began to study this topic in the early twenty-first century. Their efforts have made employment quality a policy priority, making the improvement thereof an important goal for the government. Chinese scholars working in this field have noted that the quality of employment is inevitably influenced by a nation's culture, history and unique national conditions, and have emphasized that research efforts should be "localized". Thus, most Chinese research in this field focuses on the elements, indicators, and underlying factors of employment quality that are relevant to China's national conditions and current stage of economic development.

Scholars believe that the elements intrinsic to employment quality can be classified as either subjective or objective, and either "micro" or "macro". Firstly, scholars generally accept the view that the concept of employment quality is an overarching term that reflects the quality of the process whereby labor is combined with factors of production to generate income for workers. It has been pointed out that both objective working conditions and a worker's subjective appraisal of the labor process should be included when evaluating the quality of employment (Guo 2008). Second, scholars believe that the concept of employment quality often applies at both the micro and macro level. At the micro level, quality of employment includes all elements related to an individual's employment situation (from the perspective of the worker). At the macro level, quality of employment includes the dynamic state of a labor market, at either a national or regional level.

Different scholars hold different views with regards to the specific connotations of the term "employment quality", leading to disagreements regarding how relevant indicators should be developed. Take the indicators for employment quality at a macro level for instance. Some scholars have proposed that there are six aspects to understanding employment quality: sustained increase in job opportunities, ample utilization of labor resources, the role of employment in improving family life, the role of employment in enhancing one's capabilities, the possibility of re-employment after losing a job, and ample protection of workers' rights. Correspondingly, these scholars have proposed six classes of indicators: employment environment, employability, employment situation, labor renumeration, social protections, and labor relations (Lai 2011).

As for quality of employment at the micro level, some scholars have defined employment quality as a comprehensive term that encompasses the nature of work, working conditions, job security and safety, personal dignity, health and other benefits, social security, career advancements, labor-capital relations, and equal opportunities. These nine aspects are respectively evaluated through indicators that measure work quality, labor-capital relations, benefits and social security, and career advancement (Li 2003a, b). On the whole, optimization of the labor markets, growth in workers' incomes, improvements to working conditions, better social security, and the development of harmonious labor relations have become key elements in developing indicators that reflect employment quality.

As for the underlying factors, in the early days of research on employment quality, elements leftover from the planned economy era meant that levels of administrative authority, the political landscape, and job seniority had an outsized effect on employment quality. However, with the progressive development of the market economy, human and social capital have begun to play a stronger role (Lai 2011; Meng et al. 2012; Dai and Wu 2009). Yet, household registration, ownership systems, industry type and occupation continue to exert an impact on employment quality (Su 2013).

On the whole, research into the quality of employment in China is still in the initial phase. Scholars have made inroads in defining the various connotations of the term "employment quality" and establishing the academic credentials of the concept itself. However, different scholars continue to take different approaches in designing relevant indicators. In addition, due to constraints on the availability of data, there are varying levels of inadequacies with regards to how comprehensively an indicator measures employment quality. The present paper aims to provide an overview of the current state of employment quality in China through objective "micro" indicators, while also using subjective "micro" indicators to explore the main factors underlying employment quality, with due reference to the current state of research.

2.2 The Current State of Employment Quality

(1) Unemployment rates among highly educated young people remain high despite largely stable employment rates

The employment rate is one of the basic macro indicators that reflect the quality of employment. Whether or not a country or region can provide workers with greater job opportunities has become an important standard for measuring the local employment quality. China has achieved stable employment rates as the result of government policies to promote employment, greater job opportunities in the tertiary sector, progressive optimization of industry structures, and a shrinking working-age population. The average unemployment rate in urban areas was 5.1% as of end 2014, according to employment numbers for 2014 released by the National Bureau of Statistics.

At present, stable employment rates have laid a foundation for raising the quality of employment for China's workers. On the whole, we may observe the following.

First, graduates of higher education institutions now find it easier to find jobs. From the reasons for unemployment given in surveys, there was a decline from 2011 to 2013 in the proportion of graduates who are unemployed as a result of failure to find suitable employment after graduation. This figure was 17.5% in 2011 and 14.5% in 2013 (see Table 2.1), which shows that, though the number of graduates nationwide continues to increase, graduate employment has been somewhat improved under government policies that actively encourage graduate employment.

Second, long-term unemployment problems have been addressed, but short-term unemployment rates are on the rise. From survey results, it can be observed that short-term unemployment is on the rise, while long-term unemployment is on the decline. In 2013, short-term unemployment lasting six months or less was 43.1%, nine percentage points higher than in 2011 (see Table 2.2). This indicates that problems

Table 2.1 Causes of unemployment (%)

Reason for unemployment	2008 (N = 261)	2011 (N = 106)	2013 (N = 295)
Resignation/retirement	3.6	3.6	3.2
Unable to find employment after graduation	9.1	17.5	14.5
Homemakers	14.2	13.9	24.8
Unemployment due to employer-side reasons, such as bankruptcy, enterprise reform, *xiagang, neitui, maiduan gongling*, or dismissal	36.4	14.8	16.0
Personal reasons (family factors, health, resignation, etc.)	20.1	27.8	27.0
Requisition of household contracted land	1.8	1.9	3.2
Others	14.8	20.3	11.4
Total	100	100	100

Table 2.2 Duration of unemployment (% of respondents)

Duration	2008 (N = 261)	2011 (N = 106)	2013 (N = 295)
Less than six months	24.5	34.1	43.1
Six months to one year	5.9	12.3	10.9
More than one year	69.7	53.6	46.0
Total	100	100	100

Table 2.3 How likely are you to lose your job within the next six months?

Response	Number	%	Response	Number	%
Very likely	359	7.4	Very unlikely	2069	42.4
Likely	673	13.8	Unsure	181	3.7
Neither likely nor unlikely	379	7.8	Total	4877	100
Not likely	1216	24.9			

of long-term unemployment, caused by China's structural economic transition, are being gradually alleviated, while cyclical and frictional short-term unemployment caused by cyclical economic fluctuations and labor mismatches have become more common.

Third, job security remains relatively high. Unemployment surveys show that 67.3% of respondents believe that it is unlikely or impossible for them to lose their jobs over the next six months, while only 21.2% of respondents feel that they may or are likely to lose their jobs over the same period (see Table 2.3), pointing towards high levels of job security.

Fourth, the proportion of unemployed persons with higher education levels has risen. Based on the educational qualifications of the unemployed, there have been a rise in unemployed high school graduates while unemployment rates for those with secondary school qualifications or lower have been in decline. The CSS 2008 shows that 33.8% of persons with a high school education or above were unemployed. The same figure was 43% in 2011 and 57.1% in 2013. As for those with secondary vocational education or higher, 7.5% were unemployed in 2008, 12.3% in 2011, and 24.2% in 2013. This indicates that despite higher employment rates for graduates of higher education institutions as a whole, unemployment rates for educated youth remain on the rise. The sustained decline in their employment quality should be an issue of concern for society.

(2) Wage growth for low-income groups have been lower than overall wage growth, leading to a wider income gap

Renumeration is the main motivation for participating in labor. Hence, in research on employment quality, income is often an important means by which researchers measure the quality of employment. Income from labor is the main means by which workers and their families meet daily expenses. In recent years, greater equity in

Table 2.4 Monthly wages for nonfarm workers

Monthly wages (by strata)	2011			2013			Average wage growth (%)
	Mean (CNY)	Number of persons	Standard deviation	Mean (CNY)	Number of persons	Standard deviation	
Lowest	612	777	400.526	699	985	449.649	14.2
Below average	1371	457	142.613	1792	1113	229.654	30.7
Average	1912	519	140.330	2408	484	148.001	25.9
Above average	2715	542	299.396	3065	762	213.087	12.9
Highest	7372	423	9386.790	9503	777	34,378.814	28.9
Overall	2460	2718	4333.102	3292	4121	15,241.412	33.8

income distribution has become a focus for the government—establishing sound mechanisms to support growth in incomes has become a policy priority.

Survey results show that the wages of workers in nonfarm sectors has increased significantly since 2011. In particular, the per-capita monthly wage for nonfarm workers had increased to CNY 3292 in 2013 from CNY 2460 in 2011, an increase of 33.8% in nominal terms. After dividing the monthly wages of nonfarm workers into five bands, the lowest band saw a somewhat lower increase in monthly wages, at merely 14% (see Table 2.4). This suggests that government efforts to raise wages for the lowest-income group have had limited effect. Wage growth for low-income groups remain tepid.

Based on the income gap for nonfarm workers, the highest group earned an income 1205 times greater than the lowest group in 2011. In 2013, this figure had increased to 1360, pointing towards a widening income gap.

(3) It is commonplace for workers to work overtime. Manual laborers work longer than non-manual laborers.

In research on the quality of employment, working hours are the main indicator for measuring work intensity. Reasonable working hours are beneficial to the physical and mental health of workers, while also beneficial to a work-life balance. Surveys show that nonfarm workers worked an average of 8.8 h a day in 2011, with 42.6% working more than 8 h. In 2013, nonfarm workers worked 9.1 h a day, with 45.1% working more than 8 h (see Table 2.5). This shows that it is common for nonfarm workers to work overtime, which lowers employment quality to some extent.

It is particularly noteworthy that it is exceedingly common for menial laborers to work overtime. Surveys show that 30% of nonfarm workers whose work requires a certain level of skill work more than 8 h a day. On average, such workers are at their job for 8.7 h a day. Meanwhile, more than 55% of nonfarm workers whose work mainly involve manual or semi-manual labor work more than 8 h a day. On average, such workers are at their job for 9.5 h each day. Individual proprietors work for 10.2 h a day, employees at private enterprises work for 8.89 h a day, while those in *sanzi*

Table 2.5 Daily working hours for nonfarm workers

Duration	2011		2013	
	Number of persons	%	Number of Persons	%
Below 8 h	441	14.3	664	13.6
8 h	1325	43.1	2010	41.3
Above 8 h	1310	42.6	2195	45.1
Total	3076	100	4869	100

qiye work for 8.86 h a day, all higher than workers in other jobs. This shows that, at present, some workers are depending on longer working hours to increase their income, while some enterprises continue to impose longer hours on their workers. Of course, intense work and longer hours result in a lower quality of employment.

(4) Improvements to the social safety net for workers

The social security system, otherwise known as the social safety net, is a system of material compensation and assistance to citizens by the state provided by law, playing a positive role in diversifying risk, guaranteeing basic living standards and maintaining social stability. Hence, it is an important indicator in evaluating the quality of employment. Survey results show that, at present, 88.2% of nonfarm workers enjoy medical insurance coverage in 2013—the highest rate among all insurance types. Retirement endowment insurance comes second—60.6% in 2013. Coverage rates are relatively low for unemployment insurance, work injury insurance and maternity insurance, respectively 23.3, 26.1 and 17.3% in 2013. Compared to 2011, the various social insurance coverage rates for nonfarm workers have increased by varying degrees in 2013 (see Table 2.6). This indicates that workers now enjoy somewhat greater levels of social security.

(5) Protection of workers' rights is steadily improving, and formal labor contracts have become more common.

Harmonious labor relations are a main indicator for evaluating the quality of employment. In general circumstances, the number of formal labor contracts, collective bargaining by workers, and participation by workers in management affairs are indicators that reflect the state of labor relations. The Chinese General Social Survey

Table 2.6 Various forms of social security for nonfarm workers

	2011		2013	
	Number of persons	%	Number of persons	%
Old-age insurance	1586	50.9	2954	60.6
Health insurance	2646	84.8	4299	88.2
Unemployment insurance	710	23.0	1130	23.3
Work injury insurance	751	24.3	1266	26.1
Maternity insurance	422	13.7	839	17.3

(CGSS) adopted rates of formal employment contracts as the main indicator for evaluating the state of labor relations. Our survey results show that there have been steady increases in the number of formal employment contracts due to greater government efforts to promote sound employment practices and protect workers' legitimate rights and interests while duly regulating the role of the market. In 2011, only 56.5% of enterprise employees signed a formal labor contract. This figure had risen to 61.3% in 2013.

In terms of category, fixed-term labor contracts have become more common. This figure was 41.1% in 2011 and 46.6% in 2013. In 2013, 88.3% of the employees of state-owned or state-controlled[1] enterprises had signed formal labor contracts with their employers. The equivalent figures for enterprises owned by a collective, private enterprises, and *sanzi qiye*[2] are respectively 74.1, 50.1 and 89.3%. Formal labor contracts have become a more common practice for most enterprises since 2011 (see Table 2.7).

2.3 Subjective Appraisals of Employment Quality

As employment quality reflects the quality of the specific means by which labor is combined with the means of production to generate income for workers, at the micro level, subjective evaluation of one's employment situation has become an important criterion for measuring the quality of employment. There are previous studies that have treated job satisfaction as an important indicator for employment quality. In fact, their authors believe that job satisfaction has replaced wages as the main indicator for measuring employment quality (Tian and Man 2013). Our chapter will build a linear regression model with job satisfaction as the dependent variable, which will describe human capital, social capital, and institutional factors impact subjective appraisals of job satisfaction.

(1) Self-reported Job Satisfaction

We asked respondents in nonfarm work to rate, on a scale of 1–10, their job satisfaction. Respondents were asked to rate aspects such as their working environment, work intensity, work safety, income and benefits, engagement during work, as well as their satisfaction with old-age insurance and health insurance. Results show that respondents expressed greatest satisfaction with work safety (7.16), followed by health insurance (6.83) and engagement during work (6.65). In contrast, respondents were less satisfied with income and benefits (5.46) and work intensity (5.80) (see Table 2.8). On the whole, job satisfaction was slightly above average.

---

[1] i.e. where the state holds a controlling stake—Trans.

[2] *Sanzi qiye*, literally "enterprises of three funding models" include sino-foreign equity joint ventures, sino-foreign contractual joint ventures, and wholly foreign owned enterprises—Trans.

Table 2.7 Formal labor contracts by employer type (%)

	State-owned and state-controlled enterprises		Collective-owned enterprises		Private enterprises		*Sanzi Qiye*		Total	
	2011 (N = 235)	2013 (N = 409)	2011 (N = 40)	2013 (N = 85)	2011 (N = 697)	2013 (N = 1274)	2011 (N = 41)	2013 (N = 82)	2011 (N = 1013)	2013 (N = 1850)
Fixed-term contracts	56.3	64.2	53.7	56.1	34.0	39.0	62.8	67.3	41.1	46.6
Non-fixed-term contracts	25.9	22.7	13.1	15.5	9.3	9.2	23.4	19.6	13.8	12.9
Probationary contracts	1.7	0.9	2.9	1.0	0.8	1.4	4.6	2.4	1.3	1.3
Others	0	0.5	5.2	1.5	0.2	0.5	0	0	0.3	0.5
No formal contract	16.0	11.3	25.1	25.5	54.9	49.0	9.2	10.5	42.9	37.9
Not clear	0	0.5	0	0.4	0.8	1.0	0	0	0.5	0.8
Total	100	100	100	100	100	100	100	100	100	100

Table 2.8 Job satisfaction

	Job satisfaction scores										Average
	1	2	3	4	5	6	7	8	9	10	
Working environment	3.4	2.1	4.6	4.2	18.5	14.7	14.4	19.9	6.8	11.6	6.55
Work intensity	5.5	5.1	8.7	7.9	19.0	12.6	12.3	15.9	6.0	6.8	5.80
Work safety	2.6	2.0	4.2	4.6	12.5	8.7	11.5	21.7	12.4	19.8	7.16
Income and benefits	6.2	5.5	9.0	8.8	22.0	15.6	13.0	12.1	3.7	4.3	5.46
Engagement during work	2.5	1.9	3.8	4.5	17.1	14.8	15.5	21.4	9.5	8.9	6.65
Old-age insurance	5.5	2.8	4.0	4.3	15.7	11.8	14.9	20.1	7.5	13.4	6.57
Health insurance	3.8	2.6	3.5	3.8	14.3	11.4	14.4	22.3	9.4	14.5	6.83

(2) Factors Underlying Job Satisfaction

The average value of the seven aspects of job satisfaction, each with equal weight, serve as the dependent variables in our analysis. Dependent variables include the dummy variables *female* (benchmark = *male*), *north China, northeast China, east China, central China, southwest China* (benchmark = *northwest China*), *post-60s, post-70s, post-80s, post-90s* (benchmark = *post-50s or older*), *white-collar* (benchmark = *blue-collar*),[3] *public enterprise, non-public enterprise* (benchmark = *sole proprietorship*), *non-rural household registration* (benchmark = *rural household registration*), *non-migrant worker* (benchmark = *migrant worker*), *monopolistic industry* (benchmark = *non-monopolistic industry*),[4] *formal employment*

[3] Policemen, military servicemen, as well as leading officials/executives, professional/technical staff, clerks, and persons holding similar appointments in government departments, Party organs, "mass organizations", public institutions, and enterprises fall under the category of white-collar employment. Ordinary enterprise workers, service staff, production/logistics workers and persons holding similar appointments are classified as blue-collar workers.

[4] Monopolistic industries include the power, gas and water utility industries, logistics, transport, warehousing, postal services, finance, and real estate. Non-monopolistic industries include mining, manufacturing, construction, the information transmission, IT services and software sector, distribution and retail, accommodation and F&B, rental and commercial services, water conservation, environmental and public facilities management, residential services, and other service industries. Non-profit industries include scientific research, technical services and geological surveys, education, health, social security, social welfare, the cultural, sports and entertainment sector, public administration, and social organizations.

Table 2.9 Descriptive statistics

Dependent variable	Job satisfaction (dummy variable)	Mean	Standard deviation	Sample size
		6.4038	1.53507	4086
Gender	Female	0.3557	0.47880	4086
Region	Northwest China	0.1237	0.32923	4086
	Northeast China	0.0755	0.26428	4086
	East China	0.3328	0.47126	4086
	Central China	0.2841	0.45105	4086
	West China	0.1140	0.31789	4086
Age group	Post-60s	0.2252	0.41777	4086
	Post-70s	0.2960	0.45656	4086
	Post-80s	0.2978	0.45734	4086
	Post-90s	0.0890	0.28477	4086
Type of employment	White-collar	0.3141	0.46422	4086
Institutional factors	Public enterprise	0.2994	0.45805	4086
	Non-public enterprise	0.5317	0.49906	4086
	Non-rural household registration	0.3926	0.48838	4086
	Non-migrant worker	0.7952	0.40364	4086
	Monopolistic industry	0.1018	0.30239	4086
	Formal employment	0.4353	0.49586	4086
Human capital	Years of schooling	10.7165	3.67324	4086
	Vocational skills	2.5675	1.24617	4086
Social capital	Satisfaction with social life	7.17	1.768	4086

(benchmark = *informal employment*),[5] *educational attainment*, *vocational skills*,[6] *satisfaction with social life* (see Table 2.9).

On the whole, employment quality, as evaluated through job satisfaction, displays the following characteristics.

[5] We define persons in formal employment as persons employed by a registered company, registered non-profit organization, government department, or other such entities, who have signed a formal labor contract (as well as those exempt from this requirement). Persons working for such employers, but who have not signed a formal labor contract (in contravention of regulations), are defined to be "informally employed by a government-recognized employer". Self-employed persons and those in flexible forms of employment are defined as informally employed.

[6] The vocational skills variable mainly reflects the skill intensity of a worker's current job position. It is graded on a scale of 1–5. A score of 5 means that the job requires a "very high" level of professional skill, 4 means that the job requires a "higher" level of professional skill, while 3 means that the job requires "some" professional skill. A score of 2 indicates that the job is "semi-skilled" while 1 indicates that the job involves menial labor.

First, males and females report statistically significant differences in employment quality, with females expressing higher levels of job satisfaction. In terms of wage levels, income for the average woman is merely 48.7% that of the average man (CNY 5765 for men; CNY 2810 for women). However, women report far higher levels of satisfaction with work safety, work intensity, and their working environment. Women also report slightly higher levels of satisfaction with their income and job benefits. This suggests that women are more satisfied with their working conditions. A possible interpretation is that women have less demand for high wages and greater benefits.

Second, at the regional level, workers in northern and eastern China report higher levels of job satisfaction. This suggests that in these developed regions, where white-collar workers make up a larger proportion, workers enjoy greater wages, benefits, work safety, and lower work intensity, enhancing their job satisfaction.

Third, compared to those born in the 1950s, those born in the 1960s, 1970s, or 1980s report significantly lower levels of job satisfaction, suggesting that as the main labor force demographic, these groups face greater job pressures than those born in the 1950s, who are close to retirement. Meanwhile, those born in the 1990s are just beginning to enter the labor market. The somewhat lower levels of job satisfaction reported by this group are not statistically significant.

Fourth, occupation-wise, white-collar workers report significantly higher levels satisfaction with their working environment, work intensity, work safety, wages and benefits, and engagement during work than blue-collar workers. However, white-collar workers are less satisfied with old-age and health insurance, despite enjoying better benefits in these areas. This suggests that white-collar workers have stronger demands for social security than blue-collar workers.

Fifth, with regards to institutional factors, systems of ownership, household registration and industry monopolies did not exert a significant impact on the quality of employment. This indicates that, as the economic system undergoes transition, the impact of institutional factors on employment quality is weakening. However, whether or not a worker is a migrant laborer or in formal employment continues to exert a significant influence on their job satisfaction. This suggests that it remains difficult for migrant workers to integrate into local societies, which translates to lesser protections with regards to working conditions, wages and benefits, and social security. Meanwhile, the legitimate rights and interests of workers in informal employment remain vulnerable. Currently, 56.5% of workers are in informal employment, meaning that improving their employment quality will be key to raising employment quality across the board.

Sixth, with regards to human capital, educational attainment (as measured by years of education) does not have a significant effect on job satisfaction. In contrast, vocational skills levels do have a significant impact. This suggests that higher educational attainment does not necessarily lead to a good job. Given graduates continue to face difficulties in employment, lower employment quality for this group has become common. Apart from high unemployment rates, highly educated youths continue to face depressed incomes, an increase in informal employment, lower levels of job satisfaction, and an increasing mismatch between their formal education and vocational skills.

Seventh, those who report greater satisfaction with their social life report a higher level of job satisfaction, suggesting that social capital plays a strong role in improving employment quality. Social capital has an impact on the costs of seeking employment and job matching. In a competitive employment market, social capital will play a key role in enhancing employment quality (Table 2.10).

2.4 Conclusions and Policy Proposals

Through an objective description of employment quality and analysis of self-reported (subjective) levels of job satisfaction, we reach the following conclusions regarding current employment quality in China.

On one hand, employment pressures in China have eased somewhat due to a shrinking working-age population and positive government policies to promote employment. These have also led to greater job security, lower unemployment rates, and a rise in short-term unemployment accompanied by a fall in long-term unemployment. In particular, the last factor indicates that those who lose their jobs now find it easier to seek new employment. On the other hand, unemployment rates for highly educated youth remain high, leading to a fall in employment quality.

Meanwhile, wage increases have not kept pace with a growing income gap. Wage increases for low-income groups are tepid, and such workers are often forced to work longer hours to earn more. In addition, it is common for non-public enterprises to mandatorily require overtime work.

Labor-capital relations continue to improve, with formal labor contracts becoming increasingly common. Social security has also been progressively enhanced. Improvements in these aspects indicate that China has made sound progress in regulating employment practices. On the whole, China has made progress in developing sound labor market institutions. Elements of the planned economy system that created a segregated labor market are being progressively eliminated—systems of ownership, household registration, and industry monopolies no longer exert a significant impact on employment quality. However, policies that support China's huge numbers of migrant laborers should be further enhanced to ensure that they are entitled to the same employment benefits and social security as their urban counterparts. In addition, safeguards pertaining to informal employment should be expanded to adequately protect labor rights and interests for these workers.

To this end, we propose the following.

First, positive employment policies should continue to be promoted, thereby ensuring sustained improvement to the employment situation in China. While the employment situation is largely sound, a slowdown in economic growth has led to economic fluctuations. Certain regions and industries continue to face employment pressures even as the labor force continues to grow. The quality of employment available to highly educated young people remains less than ideal, even as jobs remain

Table 2.10 Linear regression coefficients (unstandardized)

	Constants	Coefficient	Standard deviation	*t*-value	Significance level
		3.470	0.166	20.925	0.000
Gender	Female	0.332	0.047	7.130	0.000
Region	Northwest China	0.214	0.103	2.067	0.039
	Northeast China	0.154	0.114	1.350	0.177
	East China	0.185	0.091	2.029	0.043
	Central China	−0.079	0.092	−0.862	0.389
	West China	0.074	0.105	0.705	0.481
Age group	Post-60s	−0.220	0.086	−2.566	0.010
	Post-70s	−0.320	0.084	−3.795	0.000
	Post-80s	−0.247	0.088	−2.792	0.005
	Post-90s	−0.142	0.108	−1.313	0.189
Type of employment	White-collar	0.269	0.063	4.284	0.000
Institutional factors	Public enterprise	0.130	0.084	1.543	0.123
	Non-public enterprise	0.288	0.064	4.496	0.000
	Non-rural household registration	0.006	0.054	0.110	0.913
	Non-migrant worker	0.120	0.056	2.122	0.034
	Monopolistic industry	−0.018	0.073	−0.239	0.811
	Formal employment	0.152	0.057	2.651	0.008
Human capital	Years of schooling	0.004	0.009	0.449	0.653
	Vocational skills	0.161	0.021	7.770	0.000
Social capital	Satisfaction with social life	0.288	0.012	23.315	0.000
Adjusted R^2		0.186			
F Statistic		47.684			
Significance level		0.000			

somewhat scarce. Hence, positive employment policies, promoting entrepreneurship, and lowering barriers faced by entrepreneurs remain important to ensuring the quality of employment.

Second, labor market mechanisms should continue to be improved, and mechanisms to allow freedom of employment should be developed. Migrant workers continue to experience a low quality of employment even though institutional factors dating from the days of planned economy no longer exert as strong an influence on China's labor markets and institutional barriers to internal migration have been progressively eliminated. There remains room for improving institutional mechanisms that promote quicker progress in urbanization, the development of open, fair, and equitable labor markets, encouraging labor migration within reasonable limits, and ensuring the integration of migrant workers into local society.

Third, the rights and interests of workers should be safeguarded through continued regulation of corporate employment practices. Although China has made sound progress in regulating corporate employment practices, as evident from the increasing numbers of workers who have signed formal labor contracts and their entitlement to social security, mandatory overtime work remains a concern and scant protection of the rights and interests of workers in informal employment remains a policy dilemma. Hence, there is a need to further strengthen regulations pertaining to corporate employment practices to protect the physical and psychological health of workers. In addition, formal labor contracts should be more widely mandated, in order to reduce the number of workers engaging in informal employment.

Fourth, skills training for workers should be more widely promoted. Our analysis in this chapter has shown that workers' skills are key to enhancing their employment quality and job satisfaction. Survey results show that, at present, merely 39.7% of workers have participated in skills training organized by their employers over the past year. Thus, skills training for workers should be a focus for the government, relevant institutions and employers.

Fifth, labor market services that provide workers with adequate protections should continue to be developed. Main channels for seeking employment have remained largely unchanged since 2011. Referrals from friends and families remain the most common channel, accounting for 53.4%. Merely 12.8% of workers are registered at a recruitment agency, while 26.9% are hired through employer-organized recruitment events or examinations. It is also noteworthy that 28.2% of workers now hunt for jobs through the internet or other forms of media, an increase of 4.3% over 2011. It is apparent that, despite inadequate employment services, the development of internet applications has had a positive effect on job seeking channels. This would create new demands on the development of future employment services.

References

Dahu, Meng, Lifeng Su, and Li Lu. 2012. Human Capital to the Result and Quality of Employment of College Graduates: Analysis of the Data Based on Questionnaire Survey [in Chinese]. *Population & Economics* 3.

Feng, Dai, and Keming Wu. 2009. Analysis on the Positive and Negative Influence of Social Capital on the Quality of Graduates' Employment [in Chinese]. *Education Science* 3.

Fuli, Guo. 2008. Guonei jiuye zhiliang yanjiu shuping: hanyi, lianghua pingjia ji yingxiang yinsu [Review of Employment Quality in China: Connotations, Quantitative Evaluation and Underlying Factors]. *China Collective Economy* 24.

Li, Junfeng. 2003a. Analysis on Differences of Employment Quality Based on Sex [in Chinese]. *Market & Demographic Analysis*, 6.

Li, Chunling. 2003b. How Does Education Give Impact on People's Economic Income: A Study on the Rate of Educational Return [in Chinese]. *Sociological Studies* 3.

Lai, Desheng. 2011. *Zhongguo laodongli shichang baogao [China Labor Market Report 2011]*. Beijing: Beijing Normal University Press.

Su, Lifeng. 2013. Woguo xin shiqi geren jiuye zhiliang yanjiu – jiyu diaocha shuju de bijiao fenxi [A Study on Individual Employment Quality in China's New Era: A Comparative Analysis Based on Survey Data. *Economist*, 7.

Suhua, Liu, and Chunmin Han. 2007. Dui jiuye zhiliang pingjia he dingqi fabu zhidu wenti tanxi [A Discussion and Analysis of the Evaluation of Employment Quality and the Regular Issuance of Relevant Statistics]. *Academic Exchange, 1.*

Tian, Yongpo, and Zihui Man. 2013. Connotation and Measurement of Employment Quality: Research Based on International Comparisons [in Chinese]. *First Resource* 4.

Wang, Junxiu. 2014. *The Theory of Social Mentality: A Macro Social Psychology [in Chinese]*. Beijing: Social Sciences Academic Press.

Chapter 3
Education, Social Mobility and Quality of Life

Chunling Li

Educational attainment has long been an important indicator in measuring the quality of life. In addition to being a fundamental human right, educational attainment opens the door towards more satisfactory jobs, higher incomes and opportunities for career development. More importantly, access to education gives the poor, the vulnerable and those of lower socioeconomic status the chance to improve their social status, thus shrinking the rich-poor divide and promoting social harmony. At the same time, the role of education in social mobility reflects the degree of social justice and the openness of the social structure. In traditional societies, where social injustice is greater and social structures are more conservative, ascribed status (for instance family background) determine a person's socioeconomic status. In contrast, in a society that is more open and equitable, achieved factors play a more important role in deciding a person's socioeconomic status.

Academic qualifications are the most important factor in determining achieved status, as education accords people with the opportunity to improve their social standing. Our present chapter will focus on an analysis of higher levels of education in China and the impact of education on individual socioeconomic status, before discussing the relationship between education and quality of life.

3.1 Developments in Education and Growth in Educational Opportunities

Education in China has developed rapidly since the founding of the People's Republic in 1949. Education attainment has grown continually, and educational opportunities have increased. Before the founding of the People's Republic, China had an extremely backward educational system, where more than 80% of the population were illiterate.

C. Li (✉)
Institute of Sociology, Chinese Academy of Social Sciences, Beijing, China

P. Li (ed.), *The Quality of Life in Contemporary China*, Research Series on the Chinese Dream and China's Development Path, https://doi.org/10.1007/978-981-16-7021-3_3

This figure was even higher in rural areas, where merely 20% of school-age children were enrolled in schools. In 1947, a total of 150,000 students were enrolled in Chinese institutions of higher learning, whereas there were 17.98 million students in secondary schools and 228.58 million students in primary schools in 1946. Given that the Chinese population numbered 470 million back then, only 3 out of 10,000 Chinese were enrolled in tertiary education; corresponding figures are 38 and 486 for secondary and primary education, respectively (Lu and Li 1991: 210). However, developments over the past 50-odd years have created far greater educational opportunities in China. Data from the sixth census conducted by the National Bureau of Statistics show that, out of every 100,000 people, 8930 had a bachelor's degree or *dazhuan*,[1] 14,032 had completed high school or secondary vocational education, 37,788 had graduated from junior secondary school, while 26,779 had completed primary school (National Bureau of Statistics 2013).

Since adopting economic reforms, China has embarked upon a new era in the development of education. The Chinese government has continually increased investment in education, and education facilities have seen sustained improvement, greatly enhancing the quality of education. However, there are still fluctuations in the increase of educational opportunities. In the first ten years of economic reform, there was a sustained decrease in opportunities for secondary school education. In 1975, 90.6% of primary school students would go on to secondary school. This figure declined to 87.7% in 1978, 75.9% in 1980 and 68.4% in 1986. There was a similar trend of decline in the secondary school graduation rate. In 1975, 60.4% of middle school students would go on to high school. This figure saw a sharp decline to 40.9% in 1978, a slight rebound in 1980 to 45.9%, but continued to decline thereafter, to 41.7% in 1985, 40.6% in 1986, 39.1% in 1987, 38.0% in 1988 and 38.3% in 1989. The main reason for the decline in opportunities for secondary school education are the marketization of the education system (leading to an increase in fees), a loosening of social control systems (a more liberal household registration system allowed rural residents to work in cities) and the marketization of the economy (the emergence of sole proprietors and non-public sector employees). These factors caused many youths from poor rural households to give up their chance at a secondary school education in exchange for work in the cities. Child labor was also common during this period.

Since 1990, more students have graduated from secondary education, while opportunities for secondary education rebounded somewhat after the implementation of relevant government initiatives, especially "Project Hope", which provided aid to children who have dropped out of school due to poverty, as well as developments in the labor market (e.g. less demand for child labor as more young migrant workers sought employment in cities). A new round of growth in educational opportunities began in 1990. Sometime around 1994, the number of students progressing from primary school to junior secondary school, and from junior secondary school to

[1] The term *dazhuan* was originally shorthand for *daxue zhuanke xueke zhuanye jiaoyu* (i.e. specialized university vocational education). After educational reforms in 1999, this became known as "higher vocational education", but the term *dazhuan* remains in common use—Trans.

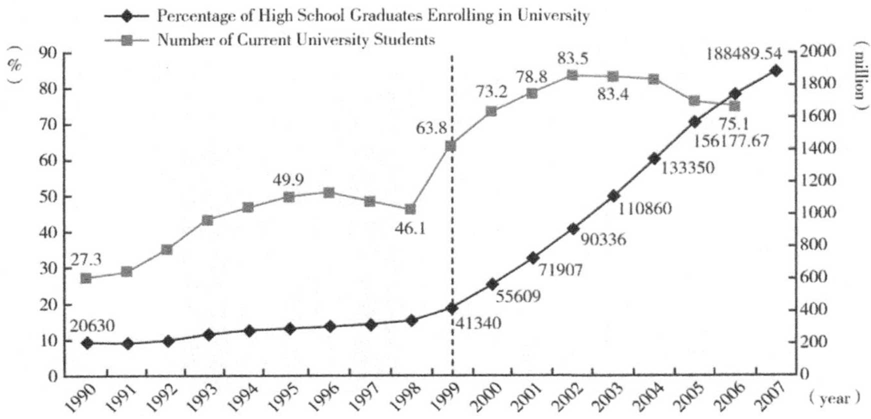

Fig. 3.1 Total number of university students and university enrollment rates for high school graduates

high school began to return to 1978 levels. Since the 1990s, growth in the supply of opportunities for basic education and secondary have picked up notably. By 2000, the nine-year compulsory education system[2] was largely instituted across the country, which effectively brought youth illiteracy rates to zero. Middle school graduation rates had exceeded 50% while high school graduation rates had reached 73.2%. Opportunities for tertiary education grew at slow rates in the 1990s. It was not until the government implemented policies for expanding university enrollment in the late 1990s that opportunities for tertiary education would grow at a breakneck pace.

In 1999, the Chinese government adopted a policy of expanding university enrollment, which led to rapid growth in enrollment rates and opportunities for tertiary education (see Fig. 3.1). Over the next five years, student enrollment in Chinese higher education institutions grew at an average annual rate of 20%. The ratio of high school graduates enrolling in university increased sharply from 46.1% in 1998 to 83.4% in 2003. In 2005 (the peak of the "enrollment expansion policy"), new student enrollment in higher education institutions was 4.7 times that in 1998, with a gross enrollment ratio of 21%. The number of students enrolled in higher education institutions that year was also 3.9 times that in 1998. Even though enrollment rates in higher education institutes have grown at a slower pace in the face of high graduate unemployment rates, enrollment growth rates have remained at around 5% since 2006.

Opportunities for tertiary education and higher secondary education have progressively grown over the past decade with the progressive implementation of the nine-year compulsory education system, higher junior secondary school graduation rates, and expanded university enrollment. Figure 3.2 shows educational attainment levels from 2004 to 2013 for persons aged six and above as a proportion of total population,

[2] Six years of primary education and three years of junior secondary education—Trans.

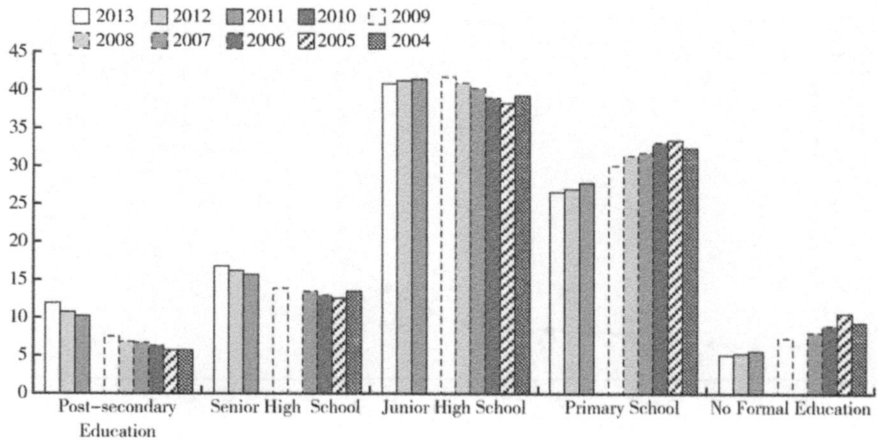

Fig. 3.2 Educational attainment (as percentage of population aged six and above)

from which we observe that opportunities for tertiary education and higher secondary education have progressively increased year after year. 11.7% of the population have received a university education as of 2013, compared to 5.7% in 2004, while 16.5% have completed high school or equivalent, compared to 13.4% in 2004. Meanwhile, there are significantly fewer people with only primary school education or are illiterate. In 2004, 32.4% of the population had received only primary school education, compared to 26.4% in 2013. Illiteracy rates have declined from 9.2% in 2004 to 5% in 2013.

Those born in the 1980s and 1990s are the biggest beneficiaries of increased educational opportunities. Government education policies and the size and quality of China's educational institutions have changed greatly over the years, meaning that every generation in China faces radically different educational opportunities. The means and likelihood of pursuing tertiary education differ greatly across generations. Based on data collected from the Chinese General Social Survey (CGSS) carried out by the China Academy of Social Sciences (CASS) in 2006, 2008, 2011 and 2013, we compare enrollment rates for different age groups during different periods, as shown in Fig. 3.3. The data clearly indicates that those born in later periods attain a higher level of education. In fact, nearly 99% of those born between 1980 and 1999 have received primary school education. Of primary school graduates, 88.9% of those born in the 1980s and 97.6% of those born in the following decade have had the opportunity for junior secondary education, respectively 14 and 22% higher than for those born in the 1970s. Of junior secondary school graduates, 53.6% of those born in the 1980s and 70.5% of those born in the 1990s have progressed to higher secondary education (high school or secondary vocational education), respectively 14 and 31% higher than those born in the 1970s. Of higher secondary education graduates, 50.8% of those born in the 1980s have progressed to university level, approximately 4% higher than those born in the 1970s. No data are yet available for those born in the

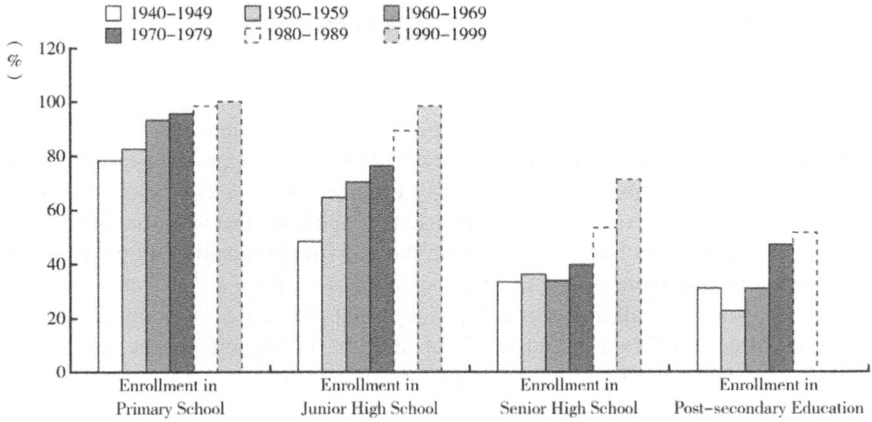

Fig. 3.3 Enrollment rates by year of birth

1990s, a significant portion of whom have yet to take college entrance examinations. However, it can be safely predicted that their university enrollment rates will be even higher.

Figure 3.4 compares gross university enrollment rates for different age groups. These figures show a clear and sustained increase in university enrollment, particularly over the past two decades. Where merely 3.9% of those born in the 1940s, 4.2% of those born in the 1950s, and 6.8% of those born in the 1960s made it to university, the college enrollment rate for those born in the 1970s was nearly twice that of those born in the 1960s. The university enrollment rate increased by 10.6% for those born in the 1980s, reaching 23.8%. Gross university enrollment rates for those born in the 1990s are even higher, where one third of age-appropriate students have enrolled in university.

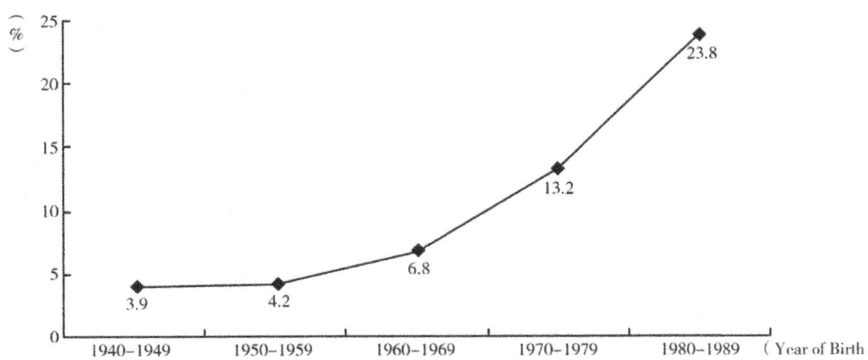

Fig. 3.4 Gross university enrollment rates by age group

3.2 Urban–Rural and Social Status Inequities in Educational Opportunities

The rapid growth in educational opportunities represented a chance to achieve greater educational equity. The greater availability of educational opportunities meant that disadvantaged social groups now had more opportunities for education. However, inequity in educational opportunities persisted though less affluent social groups and disadvantaged children had benefited from the growth in educational opportunities. When respondents were asked to rate the fairness of the *gaokao*[3] and the nine-year compulsory education system, 15.4% felt that the *gaokao* was unfair, another 15.7% reported that "it was hard to say". In addition, 11.3% felt that the compulsory education system was unfair, while another 5.4% reported that "it was hard to say".

Meanwhile, it was more likely for a highly educated respondent to believe that the *gaokao* was unfair. Per their highest level of educational attainment, 25.6% of those with postgraduate degrees, 28.0% of those with bachelor's degrees, and 27.4% of those with higher vocational education believed that the *gaokao* was unfair, compared to only 18.8% of those with secondary vocational education, 13.6% of those with junior secondary education, and 9.7% of those with primary school education. One possible interpretation is that those who have had the opportunity for tertiary education, particularly those who came from less affluent backgrounds, had a more personal experience of the unfairness of the academic rat race.

Inequity in educational opportunities is most prominently manifested between urban and rural areas, and between social groups. Table 3.1 lists the ratio distribution of those from those born in the 1980s who have graduated or dropped out from school, broken down by factors such as their fathers' occupation (as a proxy for social background), their fathers' household registration (indicating whether they are urban or rural residents). The overall ratio distributions of their fathers' occupation and household registration are included as a reference.

If educational opportunities were truly equal, graduation and drop-out rates for persons from any particular social background should be equal to overall rates, i.e. a ratio of 1. A ratio other than 1 would indicate the presence of educational inequity. The size of the disparity would indicate the level of inequity. This comparative analysis would reflect the state of educational inequity between urban/rural areas and social groups.

Although most school-age children have received primary education, a small proportion (1.8%) of those born in the 1980s have never received any formal education. The vast majority (92.2%) hail from rural villages, while only 7.8% are from urban localities. By family background, 89.1% come from farming families, while 10.1% are children of workers. By and large, they remain less well-off and lack the opportunity for upward social mobility.

The situation is largely similar for those who failed to progress to junior secondary education. In terms of household registration, 95.2% are rural residents while 4.8%

[3] i.e. the National College Entrance Examination, more often known as *gaokao* even in English publications—Trans.

Table 3.1 Graduation and drop-out rates by family background for people born in the 1980s (%)

Father's occupation (when respondent is 16 years old)	Enrolled in Primary Education		Progressed from Primary to Junior Secondary Education		Progressed from Junior Secondary to Higher Secondary Education		Progressed from Higher Secondary Education to Tertiary Education		Overall University Enrollment (population average)	University Enrollment Ratio (vs. population average)
	No	Yes	No	Yes	No	Yes	No	Yes		
Managerial executive	0	3.1	0	3.3	0.9	5.6	1.4	8.4	3.0	2.8
Professional	0	3.4	0	3.8	1.1	6.3	1.7	8.8	3.3	2.7
Clerk	0	8.7	3.4	9.3	4.0	14.1	6.6	15.4	8.6	1.8
Individual proprietor/Self-employed	0	7.5	3.7	7.9	5.9	9.7	6.5	10.3	7.4	1.4
Worker	10.9	24.0	11.1	25.6	21.4	29.5	22.5	28.0	23.8	1.2
Farmer	89.1	53.3	81.7	50.1	66.7	34.8	61.3	29.1	53.9	0.5
Total	100	100	100	100	100	100	100	100	100	–

Note Upper secondary education includes high school and secondary vocational education. Tertiary education includes bachelor's degrees and higher vocational education are urban residents. 81.7% hail from farming backgrounds, 11.1% are children of workers, while 7.1% have parents who are clerks, individual proprietors, or self-employed persons. Those born in the 1980s largely reached junior secondary age when the government was strongly promoting the nine-year compulsory education system, which led to a rapid rise in junior secondary school enrollment. However, nearly 1 out of 10 of those born in the 1980s failed to progress to junior secondary education. Much akin to those without formal education, most will probably be deprived of the chance for upward social mobility.

The path from junior secondary education to higher secondary education is a key node in the "streaming" of educational opportunities. Close to half (47.3%) of those born in the 1980s would not go above junior secondary level, of which 89.8% come from rural villages. Two thirds, or 66.7% come from farming families while another 21.4% are children of workers. Only a tiny proportion of children from other family backgrounds would miss out on the chance to advance to higher secondary education. These proportions are largely similar to those who have received only primary education or no formal education. Most are from farming households while a minority are the children of workers.

Around half of those in upper secondary education would fail to progress to tertiary education. On the whole, after "successive rounds of elimination", less than one out of four of those born in the 1980s would go on to university. If we examine the family backgrounds of those who successfully move on to the next stage of education, from primary school all the way to university, we find that the proportion of those from "higher" family backgrounds increases at each stage. Those from white-collar families or urban families make up an ever-increasing proportion, while the reverse is true for those from rural families. Table 3.1 lists the university enrollment rates for those in the 1980s from various family backgrounds as a ratio of the overall population, which clearly shows that children of professionals or managerial executives had a clear advantage in tertiary educational opportunities. The ratio of such students that attended university is approximately thrice that of their ratio in the overall population. Children of those who worked as clerks also had an advantage, where their university enrollment ratio was roughly twice their population ratio. University enrollment rates for children of workers, individual proprietors, or self-employed persons were largely around population averages, indicating neither a clear advantage or disadvantage. However, children from rural backgrounds were clearly disadvantaged—they were far less likely to attend university.

By comparing the backgrounds of students who succeed or fail in progressing to the next stage of education, we find that educational inequity persists despite the enormous growth in educational opportunities for those born in 1980–1989. There is a wide urban–rural divide in this respect, where children of professionals or managerial executives had a clear advantage, while those from rural backgrounds had the least. Of those born in the 1980s, 2% had never received formal education, while 11% have had only primary education. They hail from less affluent backgrounds, remain part of the social underclass, and their children are extremely likely to remain so. The vast majority are rural residents, as are most who fail to progress from junior secondary to upper secondary education, and from upper secondary to tertiary education. Clearly, inequity in educational opportunity is most prominently manifested in the urban–rural educational gap. The urban–rural educational gap is both the result of the urban–rural socioeconomic divide and a factor for its intensification. In addition, the effects of inter-generational transmission mean that the urban–rural educational gap will contribute towards a sustained urban–rural socioeconomic divide.

Meanwhile, inequity in the distribution of educational opportunities is not only reflected in progress to higher levels of education, but also in the different paths selected by students in the same stage of education. Students in tertiary education find that the value of a degree varies greatly from one university to another, leading to great disparities in the returns on education. Similarly, students at the upper secondary level have to choose either to enroll in high school or higher vocational education, a choice that will have a great impact on their future personal development. Students from more affluent backgrounds or urban families have a greater opportunity to enroll in a good school or otherwise receive a quality education than those from less affluent backgrounds or rural families. This implies that social status plays a role in educational inequity even for students at the same stage of education. Table 3.2 breaks down the family backgrounds of students born in the 1980s who have progressed to upper secondary and tertiary education.

First, we compare differences between the two types of upper secondary education. With ever-greater progress towards universal compulsory nine-year education, a student's choice of upper secondary education (or whether or not to enroll in upper secondary education) has become a key node in "educational stratification". After completing junior secondary education, some students would find themselves facing options that would very likely determine their personal future. Three options are presented to these students: give up the chance for further education, high school, or enroll in secondary vocational education. A little less than half, or 47.3% of those who graduated or dropped out from junior secondary school would not go on to further education, of which 89.8% come from rural villages. In terms of family background, 66.7% are from farming families, 21.4% are children of workers, while those from other backgrounds make up a tiny minority. More than half, or 52.7% will progress to upper secondary education, either to high school or vocational education. A comparison of their family backgrounds can be found in Table 3.2.

It is more common for children of professionals and managerial executives to opt for high school, where their share of high school enrollment is twice their population proportion. Their share of enrollment in vocational education is only marginally higher than their population proportion. We can thus observe that, that the upper secondary stage, students are "streamed" into three categories. Children from more affluent family backgrounds (children of managerial executives and mid/high-level professionals) often opt for high school and move on to university. Children from less affluent backgrounds (children of clerks, low-level professional, agricultural technicians) often opt for vocational education as their grades make it less likely that they can enter university, whereas a vocational education would allow them to join the labor force. Students from rural families are likely to forgo further education if their grades are inadequate for university entrance. Most would join the labor force and become migrant workers.

Table 3.2 Family backgrounds of students born in the 1980s by type of tertiary/upper secondary education

Father's occupation (when respondent is 16 years old)	Higher Vocational Education		Bachelor's Degree		High School		Secondary Vocational Education	
	Proportion	Odds ratios compared to population average	Proportion	Odds ratios compared to population average	Proportion	Odds ratios compared to population average	Proportion	Odds ratios compared to population average
Managerial executive	3.5	1.2	13.6	4.5	7.4	2.5	3.4	1.1
Professional	7.4	2.2	10.6	3.9	6.9	2.1	5.6	1.7
Clerk	16.4	2.3	14.7	1.7	13.8	1.6	14.7	1.7
Individual proprietor/Self-employed	10.3	1.4	10.2	1.4	10.5	1.4	8.7	1.2
Worker	31.9	1.3	23.4	1.0	24.5	1.0	34.5	1.5
Farmer	30.5	0.6	27.5	0.5	36.9	0.7	33.1	0.6
Total	100	–	100	–	100	–	100	–
Father's occupation (when respondent is 16 years old)	Higher Vocational Education	Bachelor's Degree	High School	Secondary Vocational Education				
	Proportion	Odds ratios compared to population average	Proportion	Odds ratios compared to population average	Proportion	Odds ratios compared to population average	Proportion	Odds ratios compared to population average
Rural household registration	35.9	0.5	27.4	0.4	52.0	0.7	41.3	0.6
Urban household registration	64.1	2.2	72.6	2.4	48.0	1.6	58.7	2.0
Total	100	–	100	–	100	–	100	–

As for tertiary education, the family backgrounds of students pursuing bachelor's degrees or higher vocational education are largely similar. As summarized in Table 3.2, children of managerial executives or professionals enroll in university at respectively 4.5 times and 3.9 times of average rates, which indicates that these groups have a clear advantage in the competition for university places. Students from other social backgrounds, with the exception of those from rural households, do not enjoy a clear advantage or disadvantage in opportunities for tertiary education, with enrollment rates largely equal to overall rates. Students from rural backgrounds are clearly disadvantaged in the competition for tertiary education, where their share of enrollment in higher vocational (and tertiary education) is only half their population proportion. Enrollment in higher vocational is somewhat higher by the same measure.

These data show that China has to devote greater efforts to addressing educational inequality despite great achievements in developing its education system. China has instituted several educational reforms after the *gaokao* system was reinstated in 1977, of which a key focus was the establishment or improvement of institutional structures—a progressively stricter examination system at each stage of education and a sound school classification system (e.g. the identification of "key" schools). Although the system has been successful in selecting and training many talented students, its outcomes have not always been fair. A rigid examination system may provide a channel for equal competition, since "all are equal in the eyes of grades", but actual competition cannot be said to be fair. Advantaged groups with greater resources will use various means to ensure greater educational opportunities for their children to ensure their success amidst fierce competition, while disadvantaged groups, especially the children of farmers, are in a weaker position, and are more likely to be "eliminated" over successive rounds of examinations. Such competition makes it likely that social inequality and urban–rural divides are passed on or even intensified in the next generation, which then leads to even greater social inequity.

In order to prevent competitive examinations from resulting in greater social inequity, there is a need to adopt measures to support disadvantaged groups in the competition for educational opportunities, lower their education expenditures and their risk of dropping out from school, enhance their returns on education, and motivate their educational pursuits. This would lead to lesser urban–rural and class disparities in educational opportunities, allowing education to play a greater role in promoting social equity.

3.3 Enrollment Expansion and Graduate Unemployment

The tertiary education enrollment expansion policy made it far more likely that those born in the 1980s and 1990s would progress to university. However, greater university enrollment created another problem—university graduates are facing greater employment difficulties, as manifested in greater rates of unemployment and inadequate employment. Since the government implemented the university enrollment expansion policy in 1999, tertiary education has expanded rapidly in China. The

Table 3.3 Gender distribution and marital status among NEETs by educational attainment (%)

Education	Gender			Marital status		
	Male	Female	Total	Single	Married	Total
Primary or below	12.2	87.8	100	7.5	92.5	100
Junior Secondary	17.0	83.0	100	19.6	80.4	100
Upper Secondary	24.7	75.3	100	27.7	72.3	100
Tertiary	43.4	56.6	100	52.0	48.0	100

number of university graduates has increased year by year, reaching 7.27 million in 2014 and 7.49 million in 2015, a 2.2-fold increase over the past decade and more than eight times greater than in 1998 (before the enrollment expansion policy was launched). This rapid growth in the number of university graduates led to a rise in graduate unemployment due to a lack of suitable jobs. According to the CGSS 2013, around one in five (20.4%) among those aged 16 to 34 were neither in education nor employment (otherwise known as "NEETs"). While most NEETs had a lower level of educational attainment, 17.4% of them (roughly 13.92 million) were university graduates. These graduate NEETs form a social group that stands out from the crowd.

The population structure of graduate NEETs is significantly different from that of less-educated NEETs, as clearly indicated in the data summarized in Tables 3.1, 3.2 and 3.3. Broken down by gender, 80% of NEETs with higher secondary education or lower are females. In terms of marital status, 80% of these NEETs are married. This suggests that less-educated NEETs are mostly married women of childbearing age. In contrast, females and married persons make up a far lower proportion of graduate NEETs. Although there are more female than male NEETs, the gender divide among graduate NEETs is far lower than their less-educated counterparts. Meanwhile, graduate NEETs are marginally more likely to be single than less-educated NEETs, which suggests that married females of childbearing age are not a dominant subset of the graduate NEET population.

Graduate NEETs often reside in different areas than their less-educated counterparts. The latter are mostly concentrated in villages and the county seats and small towns nearby, and are known as "*xianyu shehui qingnian*".[4] Conversely, graduate NEETs are more often found in "second-tier" cities, followed by "third-tier" and "fourth-tier" cities.[5] A recent dilemma facing university graduates is whether to "flee *from* Beijing, Shanghai and Guangzhou" or "flee *to* Beijing, Shanghai and

[4] Literally "county society youths", meaning rural youth who are not in formal education. In China, the term *shehui qingnian* (society youth) has many connotations, but in general, it refers to young people who are out of school and instead being "educated by society". The term may also sometimes have derogatory undertones, depending on the context—Trans.

[5] In China, it is common to (loosely) classify cities into four "tiers". First-tier cities would be those that are largest and most developed, while second-tier cities are less so, and so on. However, there is no universal definition for each tier, leading to a measure of ambiguity in the actual use of such rankings—Trans.

Table 3.4 Proportionate size of the NEET population in various urban localities (%)

Educational attainment	First-tier cities	Second-tier cities	Third-, fourth-tier cities	Towns	Villages	Total
Primary education or below	3.7	13.0	11.6	27.3	44.4	100
Junior secondary	3.8	9.9	7.9	24.1	54.4	100
Higher secondary	1.1	13.8	37.9	17.3	29.9	100
Tertiary	6.3	33.6	25.0	17.9	17.2	100

Guangzhou".[6] There are greater employment opportunities in large cities, but housing prices are formidable. Housing and daily costs of living are lower in smaller cities, but employment opportunities are scarcer, with less room for career advancement. Some graduates that "fled Beijing, Shanghai and Guangzhou" have become NEETs due to a lack of suitable job opportunities in smaller cities.

Meanwhile, the population of NEETs without tertiary education also differs across regions. There are fewer NEETs with higher secondary education in first-tier cities, which suggests that high school graduates find it easier to find jobs in these localities. However, there are large numbers of NEETs in third- and fourth-tier cities, which suggests a lack of suitable employment in these places. NEETs with junior secondary education or lower are mostly found in villages and small towns. Despite being generally unwilling to take up farming, nonfarm work is scarce in these regions (see Table 3.4).

Reasons for unemployment also differ between graduate and non-graduate NEETs, as compared in Table 3.5. Very few non-graduate NEETs have never been employed, as most have been in one form or another of unemployment after leaving school. Most report that they are unemployed for "family reasons". In other words, most are married females who choose not to work due to pregnancy or childcare. Far fewer graduate NEETs are unemployed for "family reasons", with greater numbers reporting that they have "never been employed". In fact, more than one in four graduate NEETs have never worked since graduation. Most rely on their parents for financial support, and are typical examples of the *kenlaozu*.[7] Their parents continue to support them financially despite having spent large sums on their education.

Regardless of their level educational attainment, it is highly uncommon for NEETs to be unemployed due to employer-side reasons (bankruptcy, retrenchment, enterprise reform, or *xiagang*). They are far more likely to have voluntarily resigned from their jobs, especially among graduate NEETs, where around two out of five are unemployed for this reason. In other words, this suggests that a lack of job positions is

[6] Here, Beijing, Shanghai and Guangzhou, commonly referred to as *beishangguang*, is used more as a generic term for meta-cities—Trans.

[7] Literally "those who feed off the old", i.e. adults who still rely on their parents—Trans.

Table 3.5 Duration and reasons for NEET unemployment (broken down by levels of educational attainment)

What is the main reason that you are unemployed? (%)				
	Primary Education and below	Junior Secondary Education	Upper Secondary Education	Tertiary Education
Never been employed	0.1	3.0	8.6	27.4
Family reasons	66.5	51.4	41.9	14.9
Voluntary resignation	30.6	27.5	33.7	39.6
Dismissal by employer	0.1	1.5	6.5	6.8
Others	2.7	16.6	9.3	11.3
Total	100	100	100	100
Are you currently looking for work? (%)				
	Primary Education and below	Junior Secondary Education	Upper Secondary Education	Tertiary Education
Currently seeking work	21.0	18.9	34.4	54.7
Preparing to start a business	0.5	4.7	6.9	6.8
Not seeking any form of work	78.5	76.3	58.7	38.5
Total	100	100	100	100
How long have you been unemployed? (years)				
	Primary Education and below	Junior Secondary Education	Upper Secondary Education	Tertiary Education
Average duration	5.99	2.81	2.07	1.4

not the main factor for the emergence of the NEET population, but rather that young people find it difficult to find suitable jobs. This is what economists often refer to as "structural unemployment", where a mismatch between workers and jobs leads to both large numbers of unfilled positions and high unemployment rates. Structural unemployment also increases the time needed for a worker to find a suitable job, which is most common among graduate NEETs. As shown in Table 3.5, our data indicates that more than half of graduate NEETs are currently seeking work, meaning that they are involuntarily unemployed due to a lack of suitable jobs. Length of unemployment is also different for graduate and non-graduate NEETs, where graduates have been unemployed for an average of 1.4 years, significantly lower than their non-graduate counterparts. This suggests that the graduate NEET population is a by-product of graduate unemployment. Because they take longer to find a satisfactory job, these graduates are forced to live off their parents instead of working.

3.4 Educational Attainment and Socioeconomic Status

As the population becomes more educated, levels of education have come to exert a greater influence on socioeconomic status and income growth. In most modern industrial societies, income levels are determined, to a certain degree, by educational attainment. Higher levels of education often lead to higher incomes, and vice versa, which becomes the main motivation behind the pursuit of education. Before the initiation of economic reforms, there was no clear link in Chinese society between educational attainment and income levels due to the impact of far-left policies and equalitarian distribution principles. However, since China adopted economic reforms that promoted the market economy, education has had a stronger impact on income levels, resulting in a sustained rise in returns on education.

The Institute of Econometrics at the Chinese Academy of Social Sciences (CASS) tracked 120,000 enterprise employees from 1981 to 1987. Their data showed that returns on education in urban areas was 2.5% in 1981 and 2.7% in 1987. This figure had increased to 3.8% by 1988 and 5.7% by 1995 (Zhao et al. 1999). The CASS Institute of Population and Labor Economics conducted a survey across five cities in 2002, which estimated that returns on education had grown to more than ten percent (China Adult Literacy Survey Team, 2002). Meanwhile, a nationwide randomized survey by the CASS Institute of Sociology in 2001 estimated that returns on education were 11.8% (Li 2003). The CGSS 2006, 2011 and 2013, also carried out by the CASS Institute of Sociology, arrived at similar estimates (respectively 11.6, 12.2 and 11.4%).

Table 3.6 lists the various rates of return on education, as estimated by the Mincer equation from CGSS 2013 data. The data reveals that returns on education at the national level were 11.4%, i.e. that each additional year of education led to a 11.4% gain in income. However, returns on education varies, depending on where one lives (urban or rural areas) and whether they are employed in the public or non-public sector. Returns on education were far greater in urban areas, where an additional year of education led to a 9.5% increase in expected income; this figure was merely 5.2% in rural areas. This urban–rural divide suggests that differences in industry structure between urban and rural areas lead to different income distribution mechanisms. Industries in urban areas are more specialized and skill intensive, which creates greater opportunities for professionals and technicians. Employees with a higher level of education command higher incomes, while those less educated will have to settle for less.

Meanwhile, levels of specialization and mechanization are low in China's agricultural industry, which remains highly dependent on low-skilled menial labor. A

Table 3.6 Returns on education in 2013 (estimated using the Mincer equation)

	National average	Urban regions	Rural regions	Public sector	Non-public sector
Constants	5.862 (0.128)	5.742 (0.164)	7.056 (0.187)	7.056 (0.269)	6.161 (0.145)
Years of Schooling	0.114 (0.003)	0.095 (0.004)	0.052 (0.005)	0.083 (0.006)	0.098 (0.004)
Male	0.642 (0.026)	0.555 (0.033)	0.848 (0.038)	0.201 (0.042)	0.695 (0.029)
Age	0.111 (0.006)	0.142 (0.008)	0.063 (0.009)	0.073 (0.013)	0.099 (0.007)
Age (squared)	−0.001 (0.000)	−0.002 (0.000)	−0.001 (0.000)	−0.001 (0.000)	−0.001 (0.000)
Adjusted R^2	0.232	0.206	0.183	0.164	0.118
N	10,206	10,206	10,206	10,206	10,206

higher level of education does not necessarily lead to higher incomes for agricultural workers. However, it is noteworthy that returns on education are progressively increasing in rural areas, despite still being far lower than in urban areas. Returns on education for rural villages were close to zero in the 1980s and 1990s. Even in 2001, returns on education were less than one percent (Li 2003). Yet, this figure had increased to 5.2% by 2013, suggesting that agricultural production has become more specialized and skill intensive. Since a greater of skill is now required in the agricultural sector, educated workers will enjoy greater economic returns from employment in rural areas, whether in agriculture or otherwise.

Furthermore, the different returns on education enjoyed by public sector employees and their non-public sector counterparts are reflective of the different levels of marketization between the public and non-public sectors. The non-public sector is more marketized, leading to greater returns on education, where an additional year of schooling leads to a 9.8% gain in income. Wage structures in the public sector retain vestiges of the planned economy. Because market mechanisms play a smaller role, returns on education are lower in the public sector, where an additional year of schooling generates only 8.3% in additional income. Returns on education will be lower in the public sector over the long run, as China's market reforms are first promoted in the non-public sector before being implemented in the public sector.

Education contributes not only towards greater incomes but also upward social mobility. Those with higher levels of education are more likely to achieve higher social status and job positions. The correlation between educational attainment and social status/job position is shown in Fig. 3.5. In our survey, respondents were asked to indicate if their relative socioeconomic status was "high", "above average", "average", "below average", or "low". Those with higher levels of educational attainment tended to have a favorable view of their social status, where those who graded themselves as "above average" had the most years of schooling (10.58 years). Those who described their social status as "low" had only 8.07 years of schooling (Fig. 3.6).

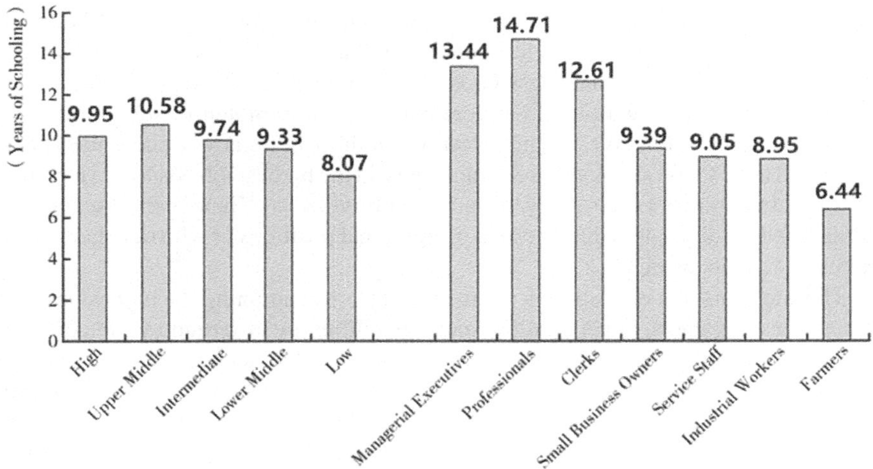

Fig. 3.5 Years of schooling by social status and employment type

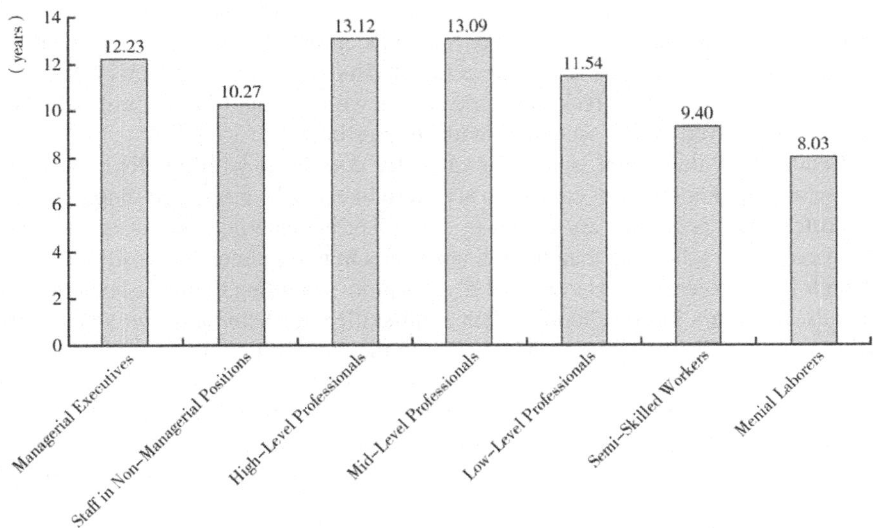

Fig. 3.6 Years of schooling by employment type

Similar trends are exhibited in employment type, where greater educational attainment leads to higher job positions. Average years of schooling is significantly greater for white-collar workers than blue-collar workers. On average, those in the three main types of white-collar employment (managerial executives, professionals, and clerks) have respectively had 13.44, 14.71, and 12.61 years of schooling. In addition, those in "medium–high" white-collar jobs (managerial executives and professionals) had more schooling than those in "lower" white-collar positions (clerks). Blue-collar workers have three to five years less schooling than their white-collar counterparts, while small business owners, service staff, and industrial workers have had only around nine years of schooling. Farmers have the least amount of schooling (only around six years). However, it is noteworthy that those of highest social status (those who rated their social status as "high") and those in the highest job positions (managerial executives) are not necessarily the most highly educated. Their years of schooling, while substantial, are less than those in professional positions or who rate their social status as "above average".

This suggests that education is not the sole factor in attaining the highest level of social or career status. Other qualities and capabilities are also required. In addition, another noteworthy phenomenon is that average levels of educational attainment are not especially high among those of higher social and career status, even though they do have more schooling than their "lower" counterparts. On average, white-collar workers who rate their social status as "above average" have close to eleven years of schooling, while those who graded their social status as "high" and "average" have approximately ten years of schooling. In other words, persons of "above average" social status are, by and large, high school graduates. Most white-collar workers have received an average of thirteen to fifteen years of schooling. The majority of those in professional positions have received tertiary education, while approximately half of managerial executives and clerks have completed tertiary education. However, nearly 50% of managerial executives and clerks, along with one out of three professionals, have only received higher secondary education or less.

Educational attainment is also significantly correlated with employment type. Those with higher levels of education are more likely to be employed in managerial or professional positions. As shown in Table 3.6, managerial executives have had two years more schooling than their counterparts in non-managerial positions. Mid- to high-level professionals have had 1.5 years more schooling than low-level professionals, 3.5 years more schooling than semi-skilled workers, and five years more schooling than menial laborers. Much akin to the distribution of educational attainment among groups of different social or career status, a significant proportion of those in managerial or professional positions have never received tertiary education, despite higher overall levels of educational attainment. One in two mid- or high-level professionals, two-thirds of low-level professionals, and three out of five managerial executives have never received tertiary education.

3.5 Conclusion

Since reform and opening up, and particularly over the past 15 years, developments in education have created greater educational opportunities in China. Illiteracy rates have declined greatly, while junior secondary education has become the universal norm. There are far greater opportunities for higher secondary and tertiary education. Those born in the 1980s or 1990s are far more likely to progress to university, leading to great improvements in population quality. Higher educational attainment has also led to greater potential for personal development and upward social mobility. With greater returns on education, education has become a greater factor in determining one's socioeconomic status, reinforcing its role in social stratification. Greater educational attainment makes it more likely that one will earn a higher income, and achieve a higher social and career status. This sufficiently reflects that achieved factors, such as education, have come to play a stronger role in social stratification, whereas ascriptive factors (e.g. family background, household registration, gender) are playing a lesser role. In other words, principles of social distribution are becoming more equitable.

However, greater educational opportunities have not led to greater equity in educational opportunities. Social status and urban–rural divides continue to contribute to inequities in educational opportunities. Children from less affluent families and rural households continue to underperform in the intense competition for educational opportunities, while more and better educational resources are available to their peers from affluent backgrounds. The problem of graduate unemployment, brought about by an expansion in university admissions, have made rural and less affluent families less willing to invest in education. This has further widened the urban–rural educational gap and narrowed their potential for development, perpetuating the current state of socioeconomic inequality.

References

Li, Chunling. 2003. How Does Education Give Impact on People's Economic Income: A Study on the Rate of Educational Return [in Chinese]. *Sociological Studies* 3.

Lu, Xueyi, and Peilin Li (eds.). 1991. *Zhongguo Shehui Fazhan Baogao [China Social Development Report]*. Shenyang: Liaoning People's Publishing House.

National Bureau of Statistics. China Statistical Yearbook. 2013. An English version is available from http://www.stats.gov.cn/english/Statisticaldata/AnnualData/

Zhao, Renwei, Li Shi, and Carl Riskin (eds.). 1999. *Zhongguo jumin shouru fenpei zai yanjiu: jingji gaige he fazhan zhong de shouru fenpei [A Restudy of Income Distribution in China: Income Distribution During Economic Reform and Development]*. Beijing: China Financial & Economical Publishing House.

Chapter 4
Environmental Consciousness and Quality of Life

Yan Cui

As greater living standards lead to a greater emphasis on the quality of life, environmental protection and environmental quality have become greater issues of public concern. Quality of life is no longer merely about basic needs such as food, clothing, housing, transport, and basic amenities. As hazy skies become a part of life in major Chinese cities, environmental factors such as air quality have become important variables for measuring the quality of life.

In recent years, China's modernization has led to intensified environmental issues. A balance between economic development and environmental protection has become a major precondition for achieving sustainable social development. However, it is undeniable that, at present, public environmental participation is somewhat low, especially in "spontaneous" environmental organizations. On the one hand, there is low public awareness of environmental issues. Most people continue to have a "free-rider" mindset with regards to environmental issues, and are not likely to take proactive actions to protect the environment.

On the other hand, because environmental oversight and means for environmental protection are currently limited, there is a lack of efficient channels for civic participation in the formulation and execution of environmental policies. In particular, members of the public rarely participate in institutionalized environmental organizations. Thus, an environmental governance structure based on civic participation can be developed only by establishing new models of social governance, exploring effective environmental governance systems, and vigorously promoting civic participation in environmental organizations.

The present chapter focuses on public environmental consciousness and the mindsets underlying response to environmental advocacy. In addition, we undertake an

Y. Cui (✉)
Institute of Sociology, Chinese Academy of Social Sciences, Beijing, China

P. Li (ed.), *The Quality of Life in Contemporary China*, Research Series on the Chinese Dream and China's Development Path, https://doi.org/10.1007/978-981-16-7021-3_4

empirical analysis of public response to environmental advocacy from the perspectives of environmental concerns, internal and external efficacy with regards to environmental issues, and inclination towards political action. We will also discuss means by which to enhance public environmental participation.

4.1 An Overview and Review of Relevant Literature

4.1.1 Existing Research on Environmental Consciousness

In existing research, there has been much debate among scholars on the environmental consciousness of the general public. Scholars have also explored the relationship between environmental consciousness and participation in environmental organizations (Wang 1999; Ren 2002: 89–113; Zhou 2008). Research reveals that there are many possible motivations behind civic participation in environmental organizations. This diversity in motivations can be attributed to the multi-dimensional nature of civic environmental consciousness.

At present, there is debate among scholars as to how exactly environmental consciousness should be defined. Existing literature suggests that there are several aspects to environmental consciousness, namely environmental knowledge, environmental attitudes, environmental values and environmental action (Hong 2006; Liu 2008). From a quantitative perspective, scholars have also proposed various scales for measuring environmental consciousness, the most representative being the New Environmental Paradigm proposed by Dunlap in 1978, and the New Ecological Paradigm proposed by a team of scholars led by van Liere in 2000. These scales have been widely used by researchers to measure environmental consciousness (Dunlap and van Liere 1978, 10–19; Dunlap et al. 2000, 425–442). In addition, the World Values Survey also includes unique tools proposed by the American scholar Inglehart for measuring environmental consciousness. In recent years, Chinese academics have also proposed methods of measurement appropriate to China's basic national conditions (Hong 2006; "Research on Systems of Indicators for Public Appraisal of Environmental Quality in China"[1] Research Group 2010).

Based upon measurements of environmental consciousness, scholars have further explored the underlying factors and have reached a certain level of consensus with regards to the following factors: (1) the impact of one's worldview (Zhong et al. 2003); (2) the impact of post-materialism[2] (Hong 2005; Hu 2006); (3) and the impact of pollution in raising environmental consciousness (Tong 2002; Liu 2008; Yan 2008). We shall explore these factors in greater detail in the following sections.

---

[1] *Zhongguo gongzhong huanjing suzhi pinggu zhibiao tixi yanjiu.* –Trans.

[2] Referring to post-materialist values as defined by Inglehart, i.e. where individuals come to value autonomy and self-expression over the material aspects of life. –Trans.

Some scholars have pointed out that one's worldview decides one's mindset, and in turn influences one's behavior patterns. With respect to environmental consciousness, we can divide the worldviews that influence civic environmental consciousness into "anthropocentrism" and "ecocentrism" (Liu 1997; Cao 2002; Bao and Wang 2003; Chen 2003). Anthropocentrism holds that the world is centered upon human beings, where nature should be in the service of humankind. From the perspective of environmental sociology, anthropocentrism places humankind at the center of the relationship between nature and human beings, where human beings are the overlords of the natural world—nature is thus an "object" that we are at liberty to conquer and remake. Guided by this moral paradigm, the interests of humankind take supreme precedence, where maximizing the interests of humankind becomes the ultimate goal of all social and economic activities. The development and utilization of nature's bounty then becomes destructive in essence, where the development of human society deviates from the laws of the natural world, eventually exceeding the carrying capacity of the natural environment and ultimately leading to environmental crises. Typical manifestations of the anthropocentric paradigm include hyper-consumerism and a commitment to "economic development at all costs".

Conversely, ecocentrism holds that human beings are merely part of the natural world and that the development of human society must be in harmony with nature. Social development must respect the carrying capacity of the environment and abide by the laws of natural ecosystems. Thus, we cannot deny that the different environmental values represented by different worldviews have an influence on environmental consciousness and in turn environmental behavior (Tong 2002).

Apart from studying environmental consciousness from the perspective of different worldviews, the American scholar Inglehart proposed in the 1970s that we may divide the ethical paradigms behind environmental consciousness into materialist and post-materialist based on one's priorities of life. Inglehart further proposed that economic development and social transformation in Western countries have led to notable changes to people's value systems. He believed that, with improved living standards, people's demands would no longer be solely focused on material enjoyment. "Materialistic" demands would be replaced by "post-materialistic" values, which are centered upon self-expression and self-actualization. "Post-materialists" disputed value systems that exclusively emphasized economic growth and material enjoyment. Introspections on development of society and the value of the individual led post-materialists to advocate that self-actualization and comprehensive improvements to the quality of life were in fact more important benchmarks for measuring happiness. On this basis, Inglehart argued that there was a close relationship between post-materialist values and environmental consciousness, and that a greater emphasis on the natural environment reflected the transition of people's needs from a lower level (e.g. material needs) to a higher level (e.g. a greater quality of life and spiritual fulfilment). Hence, environmental consciousness can be seen as a product of post-materialist values: those with post-materialist values exhibited greater levels of environmental consciousness compared to their materialist counterparts.

Furthermore, apart from exploring environmental consciousness from the perspective of abstract value systems, scholars have also focused on the concrete factors that motivate concern for the environment, which have led them to the idea of "pollution-driven environmental concern"[3] (Liu 2008). Some scholars have pointed out that the central tenet of pollution-driven environmental concern is based on environmental consciousness arising from an individual's considerations of his or her own interests. Participation in environmental protection inspired by real environmental pollution problems can be said to be "reactive", because such awareness and the corresponding environmental behavior adopted are temporary and in their own interest. Environmental participation by the general public is a result of their concerns regarding specific environmental problems and not a "greater" concern for the natural world.

In other words, public environmental consciousness may merely consist of concerns regarding environmental issues that are immediately relevant, such as air pollution, water pollution and waste pollution. When their own interests are harmed, people may seek to participate in environmental organizations on their own initiative and exhibit "defensive" environmental behavior. However, environmental problems that are at some distance from their daily lives, including global problems such as global warming and greenhouse gas emissions or ecological problems such as lesser biodiversity, deforestation and ocean pollution are paid less attention. Neither would such individuals participate in environmental organizations that focus on such issues. Hence, we may say that environmental participation inspired by immediate benefits are a form of "pollution-driven environmental action", motivated by a desire to protect one's rights, whereas environmental participation based on concern for nature and ecology are motivated by reflections upon the development of human society and the goal of maintaining harmony between human beings and nature.

4.1.2 The Inadequacies of Existing Research

As discussed above, existing research on the substance, measurement and underlying factors of environmental consciousness mainly focus on individual psychological motivations. Analysis has largely been limited to the study of social attitudes and value orientations, and lacks the ability to explain environmental behavior, particularly mass environmental participation. In terms of China's actual national conditions, environmental consciousness at the individual level does exert some influence on the actions that every citizen takes to protect the environment. However, participation in environmental organizations, which is a form of group behavior, should be studied from the perspectives of social and political participation.

As China's economy develops, environmental issues have become to affect a diverse range of interest groups. The choices of whether or not to participate in

---

[3] i.e. where an individual becomes aware of environmental issues due to pollution "in their own backyard". –Trans.

environmental organizations and whether or not to safeguard the public's environmental rights through institutional means are not only affected by individual beliefs and value systems but also by institutional factors at the state level. This is a dimension that has been neglected in existing research. Hence, our chapter proposes that, when discussing civic participation in environmental organizations, we should focus more on the attributes of social behavior and political participation inherent in environmental participation.

Political participation, broadly defined, refers to citizens influencing the enactment, decision-making process, and appraisal of the rules of state political systems and social governance mechanisms by participating in socio-political life through certain forms of organization and procedures, and thus expressing their individual or collective demands and intentions (Sun and Liao 2010). In China, participation in environmental organizations has strong attributes and undertones of political participation, making it possible to study civic environmental participation from the perspective of political participation.

Specifically, environmental participation is unlike other forms of civic engagement. For instance, participation in volunteer organizations or charity organizations is not merely determined by the participant's environmental consciousness but also influenced by particular political factors. In other words, whether or not members can truly meet their demands through environmental participation is, to a very large extent, determined by factors such as how the state defines and empowers different social groups, the ability of various interest groups to mobilize resources, the characteristics of each interest group, and their mutual recognition. An overview of environmental participation at the global level shows that environmental problems in developed Western countries is strongly political in nature. Participants in environmental organizations generally have clear political demands. When advocating environmental rights, the range of possible participant political entities is extremely wide. It is possible for any group to initiate or respond to environmental movements, regardless of class, political affiliation, occupation or gender. At the same time, when environmental problems are juxtaposed against the call for "economic development at all costs", demands regarding environmental issues may also be "generalized" into dissatisfaction with a range of social issues. Hence, when discussing the motivations behind environmental participation, we cannot ignore its underlying connotations of political participation.

In China, environmental issues are often not simply about the right to a habitable living environment. They are often related to the right to health and the right to residence. In certain regions, environmental issues are also related to ownership rights and land rights. There would be a high risk of "mass incidents" if environmental problems related to these rights are not appropriately addressed. In recent years, several environmental incidents in China have escalated into mass incidents, resulting in an adverse impact on social stability (Sun and Liao 2010). With regards to this phenomenon, Chinese scholars have debated how changes to state-society relations have contributed towards an "awakening" in public environmental consciousness and a rise in environmental movements (Hong 2007; Peng 2011).

For instance, Hong Dayong pointed out that non-governmental environmental organizations in China have become more inclined towards proactive action due to changes in the role of the state, the development of the market economy and the growth of civic society. These organizations have begun to play the role of public agents and have played a role in enhancing public environmental consciousness and addressing environmental issues (Hong 2007). In relevant research, Zhou Zhijia divided environmental participation by ordinary citizens into three categories: "informational participation", "demand-based participation", and "confrontational participation". Zhou also pointed out that citizens' participation in environmental movements reflect merely a "superficial" level of civic mindedness. One of the underlying reasons is the limited functions of non-governmental organizations in China.

Ren Liying has proposed that, at present, environmental participation by Chinese citizens is closely related to immediate personal interests, while participation at the policy level is less common (2002, 89–113). This is obviously related to the current model for policy making in China, where there are few institutional channels for civic engagement. Thus, when their rights are infringed, the public has no choice but to resort to petitioning or even other non-institutional channels to express their demands. On the whole, when studying public environmental participation, we should adequately consider the political nature of environmental participation and incorporate factors related to political participation into our research.

4.2 Environmental Satisfaction and Environmental Consciousness

4.2.1 Environmental Satisfaction

As discussed in previous studies, levels of environmental consciousness on the part of the general public are, to a certain extent, directly influenced by the quality of the surrounding environment. Higher levels of environmental pollution lead to greater public environmental consciousness. Hence, we determine levels of environmental satisfaction and environmental consciousness using data from the 2013 Chinese General Social Survey (conducted by the Institute of Sociology at the Chinese Academy of Social Sciences). The environmental module in the CGSS 2013 measures public awareness of environmental issues and public environmental consciousness.

The data indicate that public satisfaction with the environment is currently rather low in China, where nearly 25% of respondents say that they live with severe or rather severe noise and water pollution, while nearly 30% say that they live with severe or rather severe air pollution. Urban residents also express significantly lower satisfaction with the environment than rural residents. For instance, 34.8% of urban residents indicate noise pollution is "severe" or "rather severe" in their areas of residence, as compared to only 12.9% of rural residents reporting the same. Among urban residents, 38.6% report that that air pollution is "severe" or "rather severe", while the corresponding figure for rural residents is only 14.8%. Similar statistics for water pollution are respectively 30.1% and 19.5%. On the whole, data collected from surveys in various regions show that environmental satisfaction is higher among rural residents than urban residents (see Table 4.1 and Fig. 4.1).

4.2.2 Environmental Consciousness

As a consequence of rapid economic development, environmental issues have become one of the major factors that adversely impact the quality of life. As environmental problems become ever more prominent, the public has developed higher levels of environmental consciousness, leading to greater calls to protect the environment. For instance, merely 7.3% of urban and urban residents fully agree with the statement "economic development is more important to China than environmental protection", while only 3.5% of residents fully agree with the statement that "environmental protection is a government responsibility and has little to do with me". Also, nearly 75% of urban and rural residents indicate a strong willingness to join non-governmental environmental organizations. However, we should note that the "free-rider" mindset is more or less still prevalent, where 16.1% of respondents agree that "there is no need for me to care about the environment if those around me do not". In addition, nearly 70% of respondents believe that "the government should strengthen environmental protection efforts, but ordinary taxpayers should not foot the bill".

However, the data also show that there is a large urban-rural divide between Chinese citizens when it comes to environmental consciousness. For example, only 29.1% of urban residents agree that "economic development is more important to China than environmental protection", compared to 43.7% of rural residents. Only 13% of urban residents indicate that "there is no need for me to care about the environment if those around me do not", compared to 21% of rural residents. With respect to individual environmental consciousness, only 14.9% of urban residents believe that "the government should strengthen environmental protection efforts, but ordinary taxpayers should not foot the bill", compared to 24.3% of rural residents. Also, urban and rural residents differ somewhat in terms of their sense of efficacy. For instance, 50% of rural residents indicate that they do not know enough about environmental issues to comment, compared to 36% of urban residents (see Table 4.2).

Meanwhile, from the perspective of social status, levels of environmental consciousness differ greatly among different social groups. Only 20.8% of the "elite class" believe that the economy should take priority, compared to 37.4% for social groups lower on the socioeconomic ladder. In terms of individual environmental consciousness, only 7.5% of the elite class believe that "environmental protection is a government responsibility and has little to do with me", while corresponding

Table 4.1 Subjective ratings of environmental pollution (% of respondents)

	Rural regions				Urban regions			
	Noise pollution	Air pollution	Water pollution	Sanitation	Noise pollution	Air pollution	Water pollution	Sanitation
Very severe	4.5	5.1	6.9	4.1	13.6	13.0	10.2	8.5
Rather severe	8.4	9.7	12.6	14.1	21.2	25.6	19.9	21.6
Not very severe	32.3	32.0	31.5	46.2	46.2	44.2	44.7	53.6
Non-existent	54.0	52.6	48.0	34.2	18.7	16.7	23.2	15.8
Unable to say	0.8	0.7	1.0	0.9	0.3	0.5	2.1	0.5
Total	100	100	100	100	100	100	100	100

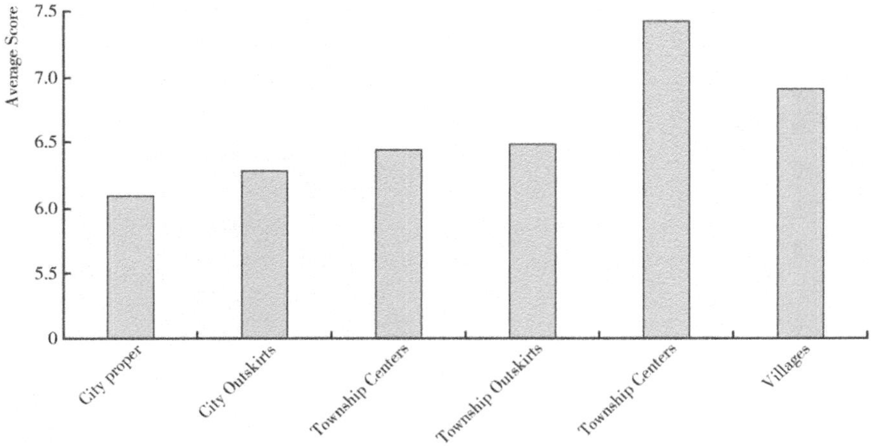

Fig. 4.1 Environmental satisfaction by locality type

figures for the middle class and "lower" social groups are 14.6 and 19.3%. As for efficacy, merely 16.6% of respondents from the elite class report a lower sense of efficacy, while this figure rises to 35.0% for the middle class and 44.2% for the "lower" social groups (see Table 4.3).

4.3 Public Environmental Participation

Past research indicates that the existence of environmental problems will lead to greater environmental consciousness and indirectly promote positive environmental behavior, whereas participation in environmental organizations is the most typical form of environmental action. Hence, we now proceed to a further analysis of public participation in environmental organizations.

Table 4.2 Public environmental consciousness (% of respondents)

	Rural areas				Urban areas				Nationwide			
	Strongly agree	Agree	Disagree	Strongly disagree	Strongly agree	Agree	Disagree	Strongly disagree	Strongly agree	Agree	Disagree	Strongly disagree
Economic development is more important to China than environmental protection	9.5	34.2	46.1	10.2	6.4	22.7	54.3	16.6	7.3	25.2	48.1	12.9
I am too busy with work, school or other daily affairs to care about environmental issues	9.5	43.2	40.8	6.6	7.5	37.1	46.0	9.4	8.1	37.2	41.4	7.6
If my peers don't care about environmental protection, neither should I	3.0	18.0	55.9	23.0	2.3	10.7	54.6	32.4	2.8	13.3	52.3	27.4
Environmental protection is a government responsibility and has little to do with me	4.5	19.8	51.9	23.8	2.6	12.3	51.5	33.7	3.5	15.2	49.0	28.5
If time allows, I am very willing to participate in environmental organizations	24.6	55.5	16.5	3.4	25.4	55.2	15.0	4.3	23.8	50.8	14.8	3.6

(continued)

Table 4.2 (continued)

	Rural areas				Urban areas				Nationwide			
	Strongly agree	Agree	Disagree	Strongly disagree	Strongly agree	Agree	Disagree	Strongly disagree	Strongly agree	Agree	Disagree	Strongly disagree
I don't know much about environmental issues and cannot comment	8.5	41.5	41.4	8.6	6.6	29.4	49.5	14.6	7.1	32.2	43.1	11.4
I have opinions on environmental issues but I don't think government departments are interested in my views	17.1	51.5	27.0	4.3	18.5	48.9	27.6	5.0	17.0	45.2	24.8	4.3
The government should strengthen environmental protection efforts, but ordinary citizens should not foot the bill	25.1	47.7	23.7	3.6	25.5	45.8	25.7	3.0	24.7	44.0	23.2	3.0
I oppose any proposal to build a chemical factory near my residence	44.1	32.9	16.2	6.8	61.6	27.3	8.7	2.4	51.8	28.0	10.7	4.2

Table 4.3 Environmental consciousness by social group (% of respondents)

	Upper class		Middle class		Workers and farmers	
	Disagree	Agree	Disagree	Agree	Disagree	Agree
Economic development is more important to China than environmental protection	79.2	20.8	71.1	28.9	62.6	37.4
I am too busy with work, school or other daily affairs to care about environmental issues	65.1	34.9	51.6	48.4	43.3	56.7
If my peers don't care about environmental protection, neither should I	93.9	6.1	86.3	13.7	82.6	17.4
Environmental protection is a government responsibility and has little to do with me	92.5	7.5	85.4	14.6	80.7	19.3
If time allows, I am very willing to participate in environmental organizations	14.6	85.4	18.3	81.7	18.4	81.6
I don't know much about environmental issues and cannot comment	83.4	16.6	65.0	35.0	55.8	44.2
I have opinions on environmental issues but I don't think government departments are interested in my views	37.7	62.3	33.2	66.8	30.3	69.7
The government should strengthen environmental protection efforts, but ordinary citizens should not foot the bill	35.8	64.2	28.8	71.2	29.4	70.6
I oppose any proposal to build a chemical factory near my residence	9.5	90.5	11.7	88.3	15.6	84.4

Note The "upper class" is defined as those who rate their social status as "high" or "above average", have completed higher vocational education or above, and who are employed in leading or professional positions in government departments, private enterprises or other similar organizations. The middle class is generally defined as those who rate their social status as "average" or "below average", have completed vocational education or above, and who are employed in the commercial service sector. Farmers and workers are defined as those who rate their social status as "low", have completed no more than junior secondary education, and who are ordinary workers or farmers

4.3.1 Research Hypotheses

Based on existing research, we propose the following research hypotheses with the level of response to environmental advocacy[4] as the casual variable to further analyze the motivations behind civic participation in environmental organizations in China.

Hypothesis 1: Higher levels of pollution in the surrounding environment lead to stronger response to environmental advocacy.

Hypothesis 2: Higher levels of environmental concern lead to stronger response by environmental advocacy.

Hypothesis 3: Higher levels of environmental efficacy lead to stronger response to environmental advocacy.

The above hypotheses follow the approaches taken by past research on participation in environmental organizations. Environmental consciousness is regarded as an endogenous motivation in our analysis of how different dimensions of environmental consciousness exert an impact on response to environmental advocacy. Hypotheses 1 and 2 are proposed to examine if citizens' response to environmental advocacy are "pollution-driven". Hypothesis 3 hopes to survey the mechanisms behind such response from the perspectives of internal and external efficacy.

Hypothesis 4: Higher levels of trust in government lead to weaker response to environmental advocacy.

Hypothesis 5: Higher levels of political consciousness lead to stronger response to environmental advocacy.

Hypothesis 6: Greater approval of extra-institutional action lead to stronger response to environmental advocacy.

Hypothesis 7: Higher levels of NIMBYism lead to stronger response to environmental advocacy.

These hypotheses aim to analyze citizens' response to environmental advocacy from the perspective of political participation. Trust in government is a major factor underlying citizens' political participation that reflects how citizens rate government efficacy. A higher level of trust in government implies that citizens have a more positive view of government efficacy. Correspondingly, trust in government is eroded

---

[4] With regards to participation in environmental organizations, the present chapter opts for the term "response to environmental advocacy" instead of "willingness to participate in environmental organizations". As revealed by a review of relevant literature, participation by Chinese citizens in environmental organizations is generally not proactive, but rather a reactive response strategy adopted when their environmental rights are infringed (Ren 2002; Hong 2008). Hence, the term "willingness to participate" is potentially misleading, whereas "response to environmental advocacy" underscores the incidental and reactive nature of such response. At a conceptual level, "response to environmental advocacy" more accurately reflects the current state of participation in civic environmental organizations. Furthermore, "response to environmental advocacy" is measured by asking a respondent if he or she will choose to participate in civic environmental organizations under specific circumstances. Here, sample bias may result if we directly measure respondents' participation in environmental organizations. In contrast to Western countries, Chinese citizens generally participate in environmental organizations to protect their environmental rights, where participation is precipitated by infringements upon such rights. "Response to environmental advocacy" is a better term, given that we are measuring the possibility of participation rather than actual participation.

when public expectations are unmet. In theory, inadequate trust in government potentially leads to self-organized civic action, where citizens take the initiative to jointly participate in or influence public affairs in an attempt to "force" the government to meet their demands. Hence, Hypothesis 4 assumes that trust in government is negatively correlated with the intensity of response to environmental advocacy.

Hypotheses 5 and 6 examine citizens' response to environmental advocacy from the perspective of political participation. As previously discussed, civic engagement by the general public, especially in environmental organizations, are not simply actions taken to maintain a clean environment. In China, the activities undertaken by many environmental organizations represent certain political demands. Research by scholars on mass incidents indicate that many environmental incidents in China have rich political undertones, given that civic engagement precipitated by environmental issues often take the form of extra-institutional action, which then leads to social instability. In contrast to other non-profit, voluntary forms of civic engagement, environmental participation in China is a form of political participation. Hence, we may posit that groups with higher levels of political consciousness are more likely to participate in environmental organizations.

Furthermore, preferences for one form of action over another reflect one's view of the effectiveness of political participation. With regards to the motivations for individual participation in environmental movements, Opp argued that individual decisions to participate are determined by the efficacy of such participation, i.e. expected efficacy of such political participation. A more positive outcome would inspire more enthusiastic participation. Meanwhile, the choice of strategy directly reflects how an individual perceives the efficacy of such participation. Hence, we posit that respondents that are more approving of extra-institutional action will have a greater response to environmental advocacy.

Hypothesis 7 studies groups that display different levels of response to environmental advocacy from the perspective of NIMBYism. As discussed above, the NIMBY mindset plays a strong role in environmental participation in China, whereas the "NIMBY phenomenon" is one of the most typical manifestations of desires to protect one's environmental rights. The "NIMBY phenomenon" mainly refers to confrontational group action taken by the public in objection to the construction of facilities that pose a pollution threat near their homes, such as waste treatment sites and chemicals factories (He 2009, 102–114; Zhang 2010). Self-interest plays a strong role in participation in environmental organizations that is highly motivated by NIMBYism. Hence, we assume that groups with different levels of NIMBYism respond differently to environmental advocacy.

4.3.2 Data and Variables

As discussed above, the data used in this chapter originate from the 2013 CASS-CGSS. The environmental issues and environmental consciousness modules of the CGSS provide variables for testing our hypotheses. In our model, citizens' response to environmental advocacy is the dependent variable. We classify respondents into three categories: persons indifferent to environmental advocacy, persons exhibiting an average level of response, and persons exhibiting enthusiastic response.

Independent variables include levels of concern for the environment, a respondent's rating of the severity of environmental pollution, internal and external efficacy (with regards to environmental issues), trust in government, political consciousness, approval of extra-institutional action, NIMBY inclinations, and their (self-reported) relative socioeconomic status. Relevant descriptive statistics are summarized in Table 4.4.

4.3.3 Analysis and Discussion

4.3.3.1 Analysis of our Model

We apply a multivariate logistic regression model for purposes of analysis. A statistical model is established by non-linear conversion of the variables and computing the proportional odds. Our model analyzes the influence of a particular independent variable on the dependent variable to test a particular research hypothesis, after controlling for other independent variables. The odds ratio in the model, exp (b), indicates that when the value of an independent variable increases by one unit, the odds ratio for the group is exp (b) times the odds ratio of the reference group. Based on the definitions of dependent variables, independent variables and control variables, we separately introduce each independent variable to test the hypotheses described in this chapter through our multivariate logistic regression model. This allows us to discuss the impact of the independent variables on the dependent variable (level of response to environmental advocacy).

As discussed above, the dependent variable represents how likely a respondent is to respond to environmental advocacy, and is divided into three levels. Those who are indifferent to environmental advocacy serve as the reference group in our regression model (see Table 4.5).

Only control variables are included in Model 1. We observe that gender does not have any impact on citizens' response to environmental advocacy. In contrast, age is significantly correlated with response to environmental advocacy, where those born in the 1970s are highly likely to respond enthusiastically. Specifically, after controlling for other variables, a person born in the 1970s is 1.392 times more likely to have an average level of response to environmental advocacy rather than being indifferent ($p < 0.001$), and 1.378 times more likely to respond enthusiastically to environmental advocacy rather than being indifferent ($p < 0.001$). Those born in the 1980s or 1990s are more likely to have an average level of response to environmental advocacy, whereas the likelihood of response is lowest among those born in the 1960s. This is to say that those born in the 1970s are most concerned about environmental issues, and are most enthusiastic in their response to environmental advocacy. The data also shows that higher educational attainment is positively correlated with the level of

Table 4.4 Descriptive statistics for model variables

Variables	Definition	%	Variables	Minimum value	Maximum value	Mean	Standard deviation
Dependent variables							
Level of response to environmental advocacy	Indifferent	26.07					
	Average	50.88					
	Enthusiastic	23.05					
Control variables							
Gender	Male	45.31	Age	17	16	41.01	13.92
	Female	54.68					
Educational attainment	No formal education	9.02					
	Primary education	21.28					
	Junior secondary education	33.12					
	Upper secondary education	12.86					
	Secondary vocational education	5.45					
	Vocational colleges	0.87					
	Higher vocational education	7.79					
	Bachelor's degree	8.61					
	Postgraduate degree	0.89					
	Others	0.05					
Residence	Urban	47.60					
	Rural	52.40					
Independent variables							
Socioeconomic status	Upper class	0.46	Levels of environmental pollution in the community	4	16	11.89	2.87
	Upper middle class	6.55	Level of environmental concern	5	20	13.50	2.33

(continued)

Table 4.4 (continued)

Variables	Definition	%	Variables	Minimum value	Maximum value	Mean	Standard deviation
	Middle class	41.27	Political consciousness	3	12	7.56	1.98
	Lower middle class	29.40	Trust in government	-4.19	3.22	0.00	1.00
	Lower class	20.64	Approval of extra-institutional action	0	2	0.18	0.53
	Unable to define	1.68	Internal efficacy	1	4	2.62	0.78
			External efficacy	1	4	2.19	0.78
			NIMBYism	1	4	1.67	0.83

Note For more details on how these variables are computed, see the module on environmental issues and environmental consciousness in the CASS-CGSS questionnaire

response to environmental advocacy [dependent variable group (1): exp (b) = 1.0, $p < 0.001$; dependent variable group (2): exp (b) = 1.3, $p < 0.001$].

On top of the existing control variables, we include views on environmental pollution and levels of environmental consciousness to form Model 2. From the model, we discover that those who feel that levels of environmental pollution in their locality of residence are high will be more likely to respond to environmental advocacy [dependent variable group (1): exp (b) = 0.97, $p < 0.05$; dependent variable group (2): exp (b) = 0.91, $p < 0.001$]. Furthermore, those who display an average or enthusiastic level of response to environmental advocacy are more likely to do so if they have high levels of environmental concern [dependent variable group (1): exp (b) = 1.172, $p < 0.001$; dependent variable group (2): exp (b) = 1.373, $p < 0.00$]. Meanwhile, it is also noteworthy that internal and external efficacy (with regards to environmental issues) have different effects on those who respond actively to environmental advocacy versus those who respond enthusiastically. For those who display an average level of response, both internal and external efficacy are positively correlated with environmental advocacy response at a statistically significant level. In contrast, only internal efficacy is a statistically significant factor for those who respond enthusiastically. In other words, external efficacy does not have a significant impact on this group. Persons in this group do not have a significantly more favorable opinion of government response to environmental problems than those who do not respond to environmental advocacy.

Trust in government and levels of political participation are included as variables in Model 3. We observe that those who display an average level of response to environmental advocacy have a significantly higher level of trust in government [exp (b) = 1.15, $p < 0.05$], whereas trust in government is not a statistically significant factor for those who respond enthusiastically or are indifferent to environmental advocacy. Meanwhile, political consciousness is also significantly stronger among those who

Table 4.5 Multivariate regression model for level of response to environmental advocacy

Dependent variable group (1) Persons with an average level of response to environmental advocacy	Model 1	Model 2	Model 3
Control variables			
Gender (Female = 0)	0.071	0.091	0.010
Born in 1960s	0.181	0.093	0.198
Born in 1970s	0.33 ***	0.312 ***	0.715 **
Born in 1980s	0.150 *	0.003	−0.071
Born in 1990s	0.259 **	0.183	0.159
Years of schooling	0.071 ***	0.028 **	0.033
Residential status (rural = 0)	0.059	0.021 *	0.029
Upper class			−0.117
Middle class			0.044
Farmers and workers			0.054
Environmental issues and levels of environmental concern			
Environmental concern		0.159 ***	0.127 ***
Pollution levels in community of residence		−0.027*	−0.050
Internal efficacy		0.271 ***	−0.187
External efficacy		−0.121 **	−0.020 **
Trust in government and proclivity towards proactive action			
Trust in government			0.138 *
Political consciousness			0.400
Degree of approval towards extra-institutional action			0.157
NIMBY inclinations			0.100 *
Self-identified social status			−0.039
Dependent Variable Group (2) Persons with an enthusiastic level of response to environmental advocacy	Model 1	Model 2	Model 3
Control variables			
Gender (0 = female)	0.059	0.078	0.051
Born in 1960s	0.132	0.083	0.500
Born in 1970s	0.321 ***	0.391 ***	0.715 ***
Born in 1980s	0.132	−0.071	−0.064
Born in 1990s	0.123	−0.029	0.199
Years of schooling	0.225 ***	0.039 ***	0.057 **
Residential status (rural = 0)	−0.210	−0.528 ***	−0.219
Upper class			0.138

(continued)

Table 4.5 (continued)

Dependent variable group (1) Persons with an average level of response to environmental advocacy	Model 1	Model 2	Model 3
Middle class			0.063
Farmers and workers			0.218 *
Environmental issues and levels of environmental concern			
Environmental concern		0.273 ***	0.292 ***
Pollution levels in community of residence		−0.089 ***	−0.057 **
Internal efficacy		0.390 **	1.189 ***
External efficacy		−0.397	−0.290
Trust in government and proclivity towards proactive action			
Trust in government			0.187
Political consciousness			0.120
Degree of approval of extra-institutional action			0.271 **
Inclination towards NIMBYism			0.273 **
Self-identified social status			−0.280 **
Adjusted R^2	0.058	0.129	0.158

Notes Persons indifferent towards environmental advocacy serve as our reference group
*** $p < 0.01$; ** $p < 0.01$; * $p < 0.05$

display an average level of response than those who are indifferent [exp (b) = 1.5, $p < 0.05$]. There is no statistically significant difference in the level of political consciousness among those who respond enthusiastically versus those who are indifferent. It is noteworthy that, in contrast, those who respond enthusiastically have significantly stronger inclinations towards NIMBYism than those who are indifferent [exp (b) = 1.31, $p < 0.01$]. This means that a respondent with a stronger NIMBY mindset is more likely to respond enthusiastically to environmental advocacy. In addition, those who respond enthusiastically also have a more favorable opinion of extra-institutional action than those who are indifferent [exp (b) = 1.31, $p < 0.01$]. This means that a respondent who has a favorable opinion of extra-institutional action is more likely to respond enthusiastically to environmental advocacy. NIMBYism has a much smaller impact on those who display an average level of response, as measured by the value of the relevant variable coefficient, which also happens to be significant only at the 5% level. Furthermore, their opinions on extra-institutional action do not have a statistically significant impact on their level of response.

Those who display an average or enthusiastic level of response to environmental advocacy are more likely to self-identify as members of "higher" social groups. This suggests that higher social status makes it more likely that one will respond to environmental advocacy [dependent variable group (1): exp (b) = 0.96, $p < 0.05$; dependent variable group (2): exp (b) = 0.75, $p < 0.01$].

Among the three models that we have considered, -2LL, AIC, and BIC reflect a relatively high goodness of fit. The various adjusted R^2 values also show that the explanatory power of our model increases as more variables are included. For instance, the adjusted R^2 for Model 3 shows that our model can explain 15.8% of the variation in our dependent variable.

4.3.3.2 A Further Discussion of our Model

Quantitative analysis shows that higher levels of educational attainment predict a greater likelihood of responding to environmental advocacy. This is consistent with past research, suggesting that education leads to stronger environmental awareness and positive environmental values. Greater levels of education also lead to stronger enthusiasm for participation in environmental organizations. Of course, from another perspective, levels of education attainment are positively correlated with socioeconomic status. Given that the middle class (and "higher classes") demand a greater quality of life than "lower classes", they are naturally more inclined to protect the environment. Meanwhile, groups of higher social status also command greater social resources, and are part of wider social networks, making them more likely to participate in social organizations. In addition, the data show that gender does not significantly influence response to environmental advocacy, whereas older groups are less likely to participate in environmental organizations.

A closer analysis of the independent variables in the model reveals that civic response to environmental advocacy in China remains focused on protecting one's environmental rights and interests. In other words, enthusiasm for participation in environmental protection groups arises only when environmental pollution becomes relatively severe in residential localities.

Of course, the data also show that concern for the environment and prioritizing environmental imperatives are an important psychological factor in determining if a person responds to environmental advocacy. This indicates that as China pushes ahead with socioeconomic development and modernization, the public will come to be more concerned with environmental quality and the ecological environment, and environmental consciousness will progressively increase. Meanwhile, environmental issues will continue to manifest during the course of China's transition to a new model of development. Hence, it is inevitable that environmental participation will become ever more common and environmental organizations ever more active.

In terms of subjective attributes, groups who respond enthusiastically to environmental advocacy differ from their counterparts in that they have stronger internal efficacy but not particularly strong external efficacy. In other words, they do not have a favorable opinion of government response to environmental issues. Their response to environmental advocacy is mainly motivated by their higher individual environmental consciousness and a stronger opinion of their individual capabilities. Furthermore, they are more approving of extra-institutional action. Our model shows that those who respond enthusiastically to environmental advocacy are more inclined to express their political demands through petitions, protests and demonstrations. At the same time, they are more strongly inclined towards NIMBYism, which has a relatively significant impact on their response to environmental advocacy. Conversely, those who display an average level of response have a milder proclivity towards political action. They are less approving of confrontational action and NIMBYism is not a strong factor in their response to environmental advocacy. However, we should note that this group has a stronger sense of individual initiative. Also, they have more a favorable opinion of government response to environmental issues (i.e. a stronger external efficacy), leading them to have greater trust in government than the two other groups.

Thus, we see that, when compared to the other groups, those who show an average level of response to environmental advocacy prefer different forms of specific action. On the one hand, in terms of how they view their political roles, this group has a higher level of civic consciousness and a certain level of concern for the environment. This stands in contrast to those indifferent towards environmental advocacy, who are generally apathetic towards social issues. On the other hand, unlike those who respond enthusiastically towards environmental advocacy, they have greater trust in government and higher external efficacy. They also do not have a strong proclivity towards confrontational political participation, instead preferring to express their views through milder means. In addition, NIMBYism does not feature strongly in their response to environmental advocacy. In other words, this group is more likely to interact in a benign manner with the government during social conflicts arising from environmental issues.

Environmental incidents in recent years have often revealed that citizens lack sufficient confidence in the state and the government. Coupled with a lack of institutional means for civic engagement, this has caused citizens to opt for reactive self-organized action, which then makes it difficult for civic organizations to interact benignly with the state. Conversely, civic organizations may adopt radical actions such as petitions, protests, and demonstrations, or even attempt to "coerce" the government through mass incidents. This may lead to an antagonistic relationship between civic organizations and the state, instead of an interactive relationship. Hence, any new model of social governance should focus on ways of enhancing trust in government and developing lawful channels for intra-institutional participation by civic organizations.

In summary, our data shows that, in today's China, there are two types of motivation behind citizens' response to environmental advocacy. The first form of motivation arises from environmental issues, which then leads to "pollution-driven, reactive environmental participation". The second form of motivation arises from a lack of adequate trust in the organs of state, where response to environmental advocacy is driven by a stronger proclivity towards political action, which suggests that "protecting one's environmental rights" is presently one of the main motivations for environmental participation in China.

4.4 Discussion and Proposals

As China's economy grows and the country becomes more urbanized, environmental issues are increasing in severity, leading to more frequent social contradictions and conflicts arising from environmental pollution that threaten current social stability. Thus, encouraging public environmental participation is one way whereby current environmental issues in China can be resolved. Meanwhile, as people become more aware of their rights, they would develop a stronger tendency to express their demands when their environmental rights are violated, whether through intra- or extra-institutional action. As Hong Dayong has pointed out, "public [environmental] participation in China is undergoing a major transition from concern for the environment to environmental action. This has manifested as a trend of rapid growth in the number of environmental disputes in China." (Hong 2008). This phenomenon requires that the government adjust its approach towards environmental oversight, promote the development of civic society, and develop innovative models for environmental governance.

It is apparent from our analysis that response to environmental advocacy in China displays the following characteristics. Firstly, response to environmental advocacy is not "proactive", i.e. not driven by environmental values, but rather "reactive", in that it takes the form of environmental action driven by self-interest. Public environmental participation is thus limited to the resolution of bread-and-butter environmental issues, instead of rising to the level of involvement in environmental policy making and participation in environmental oversight. Secondly, from the perspective of political participation, NIMBYism plays a strong role in the internal mechanisms underlying citizens' response to environmental advocacy. This reflects that the government is often slow in responding effectively to citizens' demands, as well as inadequacies in mechanisms that facilitate sound government-public interaction. Under the existing system, because channels for participation are lacking and the costs of expressing their opinions are too high, citizens are not inclined towards direct government-citizen negotiation as a means of resolution when their legitimate rights are violated. A substantial portion will instead opt for extra-institutional, confrontational action to protect their rights, which then leads to irrational mass incidents.

Of course, it has long been the case that the government plays a dominant role in environmental protection and oversight in China. The lack of institutions and structures that provide channels and opportunities for proactive public participation in environmental protection makes it impossible for the public to exert any real influence on environmental protection and oversight. Meanwhile, civic organizations also lack the resources needed to mobilize social forces. Thus, China needs to further enhance public participation capabilities and relevant institutional structures, and improve channels for public participation, to provide an institutional guarantee that public participation will be effective. With regards to promoting constructive civic environmental participation, I propose the following.

First of all, China should enhance educational efforts to develop greater civic environmental consciousness. Neither should these efforts be limited to encouraging "superficial" everyday eco-friendly behavior. Environmental literacy should not be merely limited to a "utilitarian" concern for one's surrounding environment, but should also include an understanding and awareness of the relationship between nature and humankind, as well as the development of positive environmental values. However, our analysis shows that, in today's China, response to environmental advocacy in China is primarily driven by the impact of local environmental pollution. Any effort to change the current state of affairs would require a sustained effort to nurture environmental values among citizens to enhance their overall awareness of environmental issues. This is indispensable if China is to become an eco-friendly, resource-conserving society. In other words, positive environmental values are an important precondition for proactive civic engagement. The development of positive environmental values would inspire citizens to effectively participate in the formulation of environmental policies, the appraisal of such policies, and everyday environmental oversight out of concern for the greater good instead of mere self-interest.

Secondly, China must establish and improve channels for public environmental participation. China should, through political reforms and innovations in social governance models, create the conditions necessary for civic engagement, particularly with respect to environmental issues, and expand channels for direct or indirect public participation in the enactment of environmental policies and environmental oversight. This would provide an institutional guarantee for civic engagement. The experience of Western countries show that public participation can greatly reduce the costs and increase the efficacy of environmental oversight. Thus, the objectivities of social development require that China establishes new models of environmental governance that are based on new social governance structures instead of government bureaucracy (Gong 2006). This would in turn require that China, on one hand, fully protects the lawful rights of citizens to participate in social governance and exercise their rights to participate in the political process. On the other hand, China should expand current channels for public participation, establish and improve an "environmental stakeholder" system, and strengthen the role of civic engagement in social governance. This would allow civic organizations to become guardians of the environment, which is after all the greatest of all public goods. In addition, non-profit civic organizations would be able to enhance their capabilities to organize and mobilize, ensuring that public environmental participation remains dynamic and effective over the long run.

Furthermore, environmental participation in China is currently constrained by the current political system. China has long adopted a "top-down" approach towards environmental governance and has yet to develop means for positive government-public interaction in this respect. Current systems, such as the "public hearing" model, are also unable to meet the conditions necessary for facilitating public environmental participation. Given the lack of institutional channels for participation by civic organizations in environmental governance, it is also difficult for such organizations to develop organizational capabilities that can effectively accommodate extensive participation by both social elites and the general public. This is to say that government monopoly over all environmental governance powers has directly led to a gulf between citizens and government regulators in this respect. In addition, an exclusive focus on GDP growth has led some local governments to pay only lip service to environmental issues. In turn, this leads to weak enforcement or even disregard of environmental laws, which has led to a negative impact on public environmental participation.

Meanwhile, environmental issues have become very common in China. Environmental pollution often results in extensive impacts, damaging far more than the interests of a particular individual. Thus, environmental NIMBYism often takes the form of mass action. Proper management of such incidents requires that China takes a fresh look at systems for environmental participation, in order to prevent environmental issues from evolving into social issues. Confrontational mass incidents and potential sources of social instability can be effectively prevented or eliminated only if citizens can, as a rule, participate in public administration. This in turn requires the development of systems that allow citizens to protect their rights through institutional means and orderly channels for environmental participation. These measures would effectively address the problem of information asymmetry between citizens and the government, enhance levels of mutual trust, and encourage citizens to adopt proactive and consultative forms of environmental participation instead of merely reacting to violations of their environmental rights. These measures would also strengthen the ability of civic organizations to participate in social governance, as well as enhance relevant institutions and standards for such participation.

In conclusion, we have pointed out that China needs to inspire greater environmental awareness and positive environmental values among its citizens through educational efforts. However, this is not enough to protect the legitimate environmental rights of citizens nor maximize their environmental interests, which would require the empowerment of civic environmental organizations, as well as providing institutional guarantees for public participation in environmental oversight and more channels for the public to express their environmental demands. At the same time, China can meet the demands of social development and truly enhance its environmental governance only by improving systems for environmental oversight that incorporate the forces of civic society into environmental governance practices.

References

Bao, Qingde, and Zhihong Wang. 2003. Zoujin yu zouchu zhijian: Renlei zhongxin zhuyi yanjiu shuping [To Uphold or Abandon: A Review of Research on Anthropocentrism]. *Science, Technology and Dialectics* (2).

Cao, Mingde. 2002. Ethical Transition from Human Center to Ecological Center. *Journal of Renmin University of China* (3). (in Chinese).

Chen, Jianlan. 2003. *Fei renlei zhongxin zhuyi huanjing lunlixue pipan* [A critique of non-anthropocentric environmental ethics]. In vol. 14 of *Zhexue men* [Doorway to philosophy], ed. Dunhua Zhao. Hubei: Hubei Education Publishing House.

Gong, Yingzhou. 2006. Guowai minjian huanbao zuzhi fazhan dui women de qishi [What China Can Learn from the Development of Foreign Environmental NGOs]. *Social Sciences Review* (6).
He, Yanling. 2009. 'Zhongguo shi' linbi chongtu: jiyu shijian de fenxi ["Chinese-style" NIMBY Conflicts: an Incident-Based Analysis]. *Open Times* (12).
Hong, Dayong. 2005. The Environmental Awareness of Urban Residents in China. *Jiangsu Social Sciences* (1). (in Chinese).
Hong, Dayong. 2006. Huanjing guanxin de celiang: NEP liangbiao zai Zhongguo de yunyong pinggu [Measuring Environmental Concern: Evaluation of the Application of the New Environmental Paradigm in China]. *Chinese Journal of Sociology* (5).
Hong, Dayong. 2007. *The Growing Nongovernmental Forces for Environmental Protection in China*. Beijing: China Renmin University Press. (in Chinese).
Hong, Dayong. 2008. Shi taolun gaijin zhongguo huanjing zhili de xin fangxiang [An Attempt to Discuss New Directions for Environmental Governance in China]. *Hunan Social Sciences* (3).
Hu, Liansheng. 2006. Cong wuzhi zhuyi dao hou wuzhi zhuyi – xiandai xifang shehui jiazhi linian de zhuanxiang [From Materialism to Post-materialism: The Shift in Modern Western Social Values]. *Contemporary World & Socialism* (2).
Liu, Fusen. 1997. Ziran zhongxin zhuyi shengtai lunliguan de lilun kunjing [The Theoretical Dilemmas of Eco-centric Environmental Ethics]. *Social Sciences in China* (3).
Liu, Jifeng. 2008. Daxuesheng huanjing xingwei de yingxiang yinsu fenxi [Analysis of the Factors Underlying the Environmental Behavior of University Students]. *Contemporary Youth Research* (11).
Peng, Yuanchun. 2011. Woguo huanjing xingwei yanjiu shuping [A Review of Research on Environmental Behavior in China]. *Social Science Research* (1).
Ren, Liying. 2002. Huanjing baohu zhong de gongzhong canyu [Public Participation in Environmental Protection. In *Huanjing wenti yu huanjing yishi* [Environmental Issues and Environmental Consciousness], ed. Ming Yang. Beijing: Huaxia Publishing House.
Sun, Huan, and Xiaoping Liao. 2010. Ethic Dimension of Subject in Political Participation: Also on Environmental Ethics. *Journal of Beijing Normal University (Social Sciences)* (6). (in Chinese).
Tong, Yanqi. 2002. Huanjing yishi yu huanjing baohu zhengce de quxiang – dui zhongguo liu chengshi zhengfu guanyuan he qiye zhuguan de diaocha [Environmental Consciousness and the Orientation of Environmental Protection Policies: A Survey of Government Officials and Corporate Executives in Six Chinese Cities]. In *Huanjing wenti yu huanjing yishi* [Environmental Issues and Environmental Consciousness], ed. Ming Yang. Beijing: Huaxia Publishing House.
Wang, Min. 1999. *Huanjing yishi ji ceping fangfa yanjiu* [Studies on Environmental Consciousness and Evaluation Methodology]. Beijing: China Environmental Science Press.
Yan, Bin. 2008. *Shehui bianqian zhong de huanjing fa* [Environmental Law During Social Transition]. Wuhan: Huazhong University of Science & Technology Press.
Zhang, Xianghe. 2010. NIMBY of Waste Treatment Plant and Its Solution Mechanism of Social Conflict. Ph.D. diss., Chongqing University. (in Chinese).
Zhou, Zhijia. 2008. Environmental Awareness Research: State, Crisis and Solutions [in Chinese]. *Journal of Xiamen University (Arts & Social Sciences)* (4).

Chapter 5
Civic Engagement and Quality of Life

Yan Cui

5.1 Definition and Significance of Civic Engagement

The level of civic engagement is an important indicator in measuring the quality of social development, while the factors that motivate civic participation directly reflect the quality of life enjoyed by members of a society. "Civic engagement" is an important sociological concept. Broadly defined, civic engagement refers to public action in the sociological sense, including political participation (as defined by political science). The concept of civic engagement has been noted by scholars, who have analyzed civic engagement and its underlying motivations from various different perspectives.

Civic engagement is a form of behavior where ordinary citizens influence government decisions, political systems, civic society, and political life through certain specific means. The rationality for the existence of civic engagement is based on modern social democracy, while civic engagement is also, in itself, the foundation for the existence of democratic politics and social autonomy. In different stages of socioeconomic development, the motives, action-awareness, rules and modes of participation, and goals of civic engagement are also naturally different. In existing research on civic engagement, most studies approach the topic from the perspective of political participation. In other words, civic engagement, broadly defined, refers to citizens influencing the enactment, decision-making process, and appraisal of the rules of state political systems and social governance mechanisms by participating in socio-political life through certain forms of organization and procedures, and thus expressing their individual or collective demands and intentions (Sun and Liao 2010). For instance, when defining the concept of political participation, Huntington pointed out that political participation encompasses all actions taken by members of the public to influence government policy, regardless of whether such actions were lawful, involved violence or were autonomous.

Y. Cui (✉)
Institute of Sociology, Chinese Academy of Social Sciences, Beijing, China

P. Li (ed.), *The Quality of Life in Contemporary China*, Research Series on the Chinese Dream and China's Development Path, https://doi.org/10.1007/978-981-16-7021-3_5

In addition, Huntington not only made an important distinction between autonomous participation and mobilized participation, but also between concepts such as lawful participation such as voting in elections and unlawful participation such as boycotts, protests, demonstrations (Huntington and Dominguez 1996, 188–199). In contrast, the social scientists Norman Nie and Sidney Verba have a somewhat narrower view. They saw political participation as a form of lawful action undertaken to influence the decisions of government officials. Per this definition, participative action encompasses only autonomous participation—mobilized participation is excluded, and any participative action can only be in the form of lawful action (Nie and Verba 1996, 290).

Here, I argue that research on civic engagement (particularly political participation) should focus both on lawful, regular forms of participation and their unlawful, irregular counterparts. The lawfulness of an action is directly determined by a particular socio-political structure, which means that an action may be unlawful in one society but lawful in another. Hence, the scope encompassed by the concept of civic engagement should not be constrained by political/legal institutions or ideology. In addition, civic engagement, particularly the "unlawful" sort, are direct feedback from an individual with regards to the social structure and reflect structural problems in social governance. The mode and level of civic engagement adopted by an individual is an important yardstick for measuring a country's level of social development and the maturity of its civic society, as well as an important factor in sustainable social development. Hence, comprehensive research into the mechanisms behind public civic engagement is possible only through adequate consideration of the rationality behind the existence of "unlawful actions". At present, as social development enters a new phase, promoting constructive participation by the general public in social governance and forming a positive "bottom-up" model of social governance adapted to the current trends of development are fundamental issues that China must face head-on during the course of its socioeconomic transition.

5.2 Existing Research on Civic Engagement as a Factor in Quality of Life

In terms of their fundamental social attributes, every individual in a society is a participant in social interaction. Their level and mode of social interaction are concurrently a reflection of their quality of life and means which they assess their quality of life. In other words, civic engagement is part of an individual's quality of life. On one hand, civic engagement is a bridge that connects the individual with society, i.e. the specific means by which an individual communicates with society. On the other hand, civic engagement by an individual represents social empowerment. Social empowerment refers to the level at which society guarantees the public's right to participation. This includes not only whether society grants an individual the capability to participate in social life, but also whether society has established effective institutional structures and social relations that enhance an individual's capacity for action.

In this way, an empowered society can also improve the individual's quality of life. Individuals in a society, particularly those from vulnerable groups, can take action through channels provided or tolerated by macro-institutions and effectively realize their demands through their opportunities for action. Put differently, a high-quality social life can be achieved only in an empowered society, where, to the maximum possible extent, individual agency is given free rein, individual efficacy is ensured, and individual dignity is respected. As John Friedmann argued, a highly empowered society can be viewed from the perspectives of the individual, society, and politics. Individual empowerment refers to all the knowledge, skills and experiences that enhance individual dignity and individual capacity for self-improvement; social empowerment refers to the means by which various individuals establish mutual social relations; political empowerment refers to whether the general public can obtain information and resources through participation in policy making.

Thus, in an empowered society, it is only right that a system for evaluating quality of life should encompass relevant indicators reflecting civic engagement of the public. In particular, these indicators should include ways by which the public obtains social information (i.e. accessibility to public information), their inherent rights to participate in the political process, their political consciousness, along with surveys of public willingness (and action) in participating in social affairs (Lin 2010).

As some scholars have pointed out, civic engagement is closely related to political participation. In an empowered society, citizens have wider political participation and richer political lives. In addition, political participation and civic engagement touch upon all levels of social life. Individuals also exert a multi-dimensional impact on communities, organizations and government policies through various forms of participation. Also, the individual in empowered through the process of participation, allowing members of society to better integrate into their social environments, which helps to maintain the harmony and stability of the social order.

Meanwhile, Huntington's debates with Nie and Verba can provide the following cues for research into civic engagement in China. The positive role of civic engagement in maintaining social order cannot be denied, while the view that only levels of lawful, "institutionalized" participation (for instance in labor unions, political parties, social organizations) can represent social progress and a modern civic culture is highly doubtful. In fact, it is precisely those extra-institutional, irregular forms of civic engagement that reflect the tensions between the individual and society, allowing the problem of civic engagement to be better explored from the perspective of quality of life and more reflective of the forms and qualities of social empowerment that underlie civic engagement. This then allows us to emphasize the "plasticity" and "creative potential" of macro-level social conditions during a discussion of the capacity for individual agency, which allows for a further analysis of how individual agency can be achieved through the process of civic engagement.

In conclusion, based on what we have discussed thus far, I propose the theoretical hypothesis that it is only in an empowered society that an individual would interact with society through institutional, regular forms of civic engagement to realize their individual social expectations, which then leads to a higher subjective quality of life. In contrast, in a less empowered society, the individual would more often interact with society through non-institutional, irregular forms of civic engagement to realize their social expectations, leading to a lower subjective quality of life.

5.3 Civic Engagement and the Quality of Life—An Empirical Analysis

As discussed above, as a social agent, the individual's capacity for action is influenced by both factors at the individual level and the structural constraints imposed by the macro-institutions of society. Hence, a comprehensive discussion of civic engagement must consider the channels of participation permitted within the limits of institutional structures and the choices regarding civic engagement faced by various groups, each of whom have different social attributes.

Through analysis of data from the 2013 Chinese General Social Survey, we undertake a preliminary discussion of the forms and causes of civic engagement by members of the public, with a particular focus on their political participation (e.g. participation in petitions, protests, demonstrations, and school/labor/merchants' strikes). Forms of political participation undertaken by members of the public are covered in the civic engagement module in the CGSS 2013. The module mainly asks respondents if they discuss political issues (whether with their peers or online), whether they have expressed their opinions on such issues to government authorities or through the media, and whether they have participated in petitions to government authorities, residents'/village committee elections, protests, demonstrations, or labor/school/merchants' strikes.

Based on traditional Western political theory, our chapter divides forms of political participation into three categories: (1) "general political participation", which mainly includes forms of participation that are not directed towards any particular purpose, such as discussion of political issues (whether with one's peers or online) and expression of individual opinions to government authorities or through the media; (2) "mobilized political participation", mainly referring to participation in residents'/village committee elections, which is generally government-mobilized and "passive" in nature; (3) "confrontational political participation", which mainly refers to participation in petitions to government authorities, demonstrations, protests, and labor/school/merchants' strikes. Confrontational political action is, to a certain degree, adversarial and directed towards particular purposes, involving strong initiative on the part of the individual agent. The legitimacy of such actions is often in doubt. In addition, the civic engagement module in the CGSS 2013 also measures respondents' attitudes towards political participation (their political consciousness). These data will be further analyzed in the following sections.

(1) General political participation and mobilized political participation are common among both urban and rural residents, while some are willing to engage in confrontational political participation

From our survey data, political participation by both urban and rural residents exhibit three attributes.

First, it is more common for urban and rural residents to engage in general forms of political participation. For instance, nearly 40% of residents (both urban and rural) have discussed politics with their peers and on online platforms, expressed political opinions through the media, and shared their views with government departments. However, with the notable exception of discussing politics with their peers (39.3%), most forms of political participation are relatively uncommon. In addition, intra-institutional participation, particularly in the form of mobilized political participation, has become the defining trait of political participation by urban/rural residents, where figures show that 43.1% have participated in residents'/village committee elections. Though this number may be relatively high, it does not actually reflect the level of self-initiated political participation. Such behavior is often motivated by pressure from organizations or formal norms, meaning that high participation rates do not necessarily imply a high level of social empowerment (Heberer 2005). Meanwhile, 48.5% of individuals that have never participated in resident/villager committee elections are unwilling to participate in these elections. This shows that a substantial part of the population has little enthusiasm for political participation under existing institutional arrangements and a low opinion of the efficacy of intra-institutional participation.

Secondly, members of the public are less inclined towards "non-regular" forms of political participation, especially the confrontational type. Participation in activities such as petitions to government departments is relatively rare—merely 3.5% of the population has participated in petitions, only 0.9% have participated in demonstrations, and only 0.9% have participated in school, labor or merchants' strikes.

Lastly, some groups show a greater willingness to engage in confrontational political participation. Among those who have never participated in petitions to government authorities, 23.1% indicate that they are willing to participate in such activities. Among those who have never participated in protests or demonstrations, 9.9% are willing to do so, while 7.8% of those who have never participated in labor/school/merchants' strikes are open to the possibility. This suggests that willingness to engage in extra-institutional confrontational political participation is greater among certain groups, which may potentially affect current social harmony and stability (see Table 5.1).

(2) Characteristics of Groups Willing to Participate in Confrontational Civic Engagement

The characteristics of social groups that actively participate in politics have been more comprehensively explored in traditional Western political theory. Some scholars have pointed out that levels of political participation are generally higher

Table 5.1 Inclination towards and actual participation in political action

	Persons who have previously participated in political action		Are you willing to participate in political action? (for those who have never participated)					
			Yes		No		Unable to say	
	Number	%	Number	%	Number	%	Number	%
Discussion of political issues with other persons	2854	39.3	1598	35.6	2638	58.8	251	5.6
Discussion of political issues on the internet	455	6.3	2021	29.7	4343	63.9	430	6.3
Expressing opinions via correspondence with media outlets	116	1.6	2206	31.0	4564	64.1	346	4.9
Feedback to government authorities	529	7.3	2519	37.4	3910	58.0	309	4.6
Petitions to government authorities	255	3.5	1607	23.1	5003	71.9	350	5.0
Participation in resident/village committee elections	3157	43.1	1994	47.5	2037	48.5	168	4.0
Participation in protests or demonstrations	64	0.9	710	9.9	6048	84.5	396	5.5
Participation in labor/merchants'/school strikes	63	0.9	557	7.8	6206	86.8	388	5.4

among those of higher socioeconomic status, given that these groups are more highly educated, have more resources at their disposal, and have higher levels of political consciousness, which naturally leads a stronger belief in their ability to influence government policy. Many researchers have also studied differences in the level of political participation between groups of different socioeconomic status (Huntington and Dominguez 1996, 189). However, there has been less research on confrontational and "non-regular" forms of political participation. Thus, our analysis will focus on confrontational forms of civic engagement, where we will explore in detail the attributes of groups that are more willing to engage in confrontational civic engagement, along with the factors underlying their greater inclination towards such action.

i. The diverse characteristics of confrontational groups reflect the multi-dimensional nature of contemporary social conflicts

Groups that have or are willing to engage in confrontational political participation ("confrontational groups" for short) are extremely diverse, though most of their number are of lower social status. In terms of their political affiliation, 82.8% are members of the general public, while merely 7.9% are members of the Communist Party and 8.8% are members of Communist Youth Leagues.

In terms of educational attainment, only 13.2% have received tertiary education, while the other 86.8% have received no more than upper secondary education. Rural residents constitute 47.8% while urban residents constitute 52.2%. Among this group, 71.2% have rural household registrations, while merely 28.8% have urban household registrations. Hence, we can see that China's progressive urbanization has left a substantial number of migrant workers dissatisfied with existing institutions and policies. On one hand, migrant workers are more affected by China's urban-rural divide than any other social group. On the other hand, they are generally not entitled to the same social benefits enjoyed by urban residents. In addition, although some migrant workers have a relatively strong inclination to protect their rights, their actions in this respect are often indiscriminately directed towards all forms of government authority, manifesting as mass action of an extra-institutional, confrontational nature. Hence, it is particularly important that their interests are safeguarded.

In terms of age, among confrontational groups, 21.9% are under 30 while 19.1% are between the ages of 30 and 40. The average participant in petitions was 50.4 years old, the average participant in demonstrations or protests was 37.7 years old, while the average participant in labor/school/merchants' strikes was 32.9 years old. In other words, those who are younger are more inclined towards intense forms of confrontational action.

In terms of employment status, among confrontational groups, 27% are unemployed. Among those who are currently employed, merely 19.1% work in Party organs, government departments, public institutions, or state-owned enterprises. The rest work in the non-public sector, the three most common forms of employment being private enterprises (31.0%), individual proprietors (22.8%), and no fixed employment (17.9%) (see Table 5.2).

Table 5.2 Employment structure among confrontational groups (nonfarm employment)

	Frequency	%
Party organs, government departments, people's organizations, military	33	2.5
State-owned or state-controlled enterprises	95	7.2
Public institutions (state-owned/collective)	123	9.4
Collective-owned enterprises	22	1.7
Private enterprises	407	31.0
Sanzi qiye [a]	27	2.1
Individual proprietors	299	22.8
Social service organizations	18	1.4
Autonomous organizations (residents' committees, village committees)	14	1.1
Others	20	1.5
Unemployed	235	17.9

[a] See note in Chapter II. –Trans.

In terms of social status, 23.3% of those who participate in confrontational action identify as members of "lower" social groups. In terms of income level, groups who have never participated in confrontational action have an average annual income of CNY 23,763.95, while those who have or are willing to participate in confrontational action have a significantly lower income of CNY 18,803.75.

ii. Confrontational political participation is motivated by pragmatic interests rather than ideology

The impact of socioeconomic transition on the interests of various social groups has become far more complex, while mechanisms for resolving conflicts between social groups are less than ideal. Most who participate in petitions and demonstrations do so because formal channels cannot provide effective relief when their interests have been harmed.

Figures show that, in terms of living standards, the issue most commonly cited by groups that have or are willing to participate in confrontational action is "lower living standards caused by higher prices"—69.0% of respondents feel that this has affected their quality of life. The second most commonly cited issue is that of "low household incomes and difficulties in meeting daily expenses"—51.7% of respondents feel that this has affected their quality of life. The third most commonly cited issue is that of "poor housing conditions and inability to purchase/build new housing"—42.3% of respondents feel that this has affected their quality of life (see Table 5.3).

iii. Dissatisfaction with the specific implementation of government policies is a motivation for participating in confrontational action

Those participating in confrontational action are often dissatisfied with the government. In other words, dissatisfaction with local government is a motivation for participating in confrontational action. Relevant data indicates that, on the whole, members of the public who have participated in petitions, demonstrations, or labor/school strikes have a far lower opinion of the government than those who have not. For instance, among those who have participated in demonstrations, only 27.0% believe that the government is clean, upright, and enforces strict penalties on corruption, only 28.9% are satisfied with the level of government transparency, and only 36.1% believe that government policies are faithfully and fairly enforced.

Our data indicates that those participating in confrontational action are most dissatisfied with government corruption, unfairness in the implementation of policies, and a lack of government transparency. Nearly two out of three such persons indicate that they are dissatisfied with the government, which suggests that improving government work is key to maintain social stability. One of the focuses of social development is the transformation of the role of government, where the government comes to see itself as a provider of services rather than a wielder of power (as commonly perceived). However, the trend towards government as a provider of services first and foremost requires regulation of government conduct, curbing the abuse of power, and instituting norms of integrity and incorruptibility among government officials. This would eradicate the phenomena where local residents remain dissatisfied with the government even though their localities have met performance targets. Hence,

Table 5.3 Issues encountered over the past year by each group

Issues encountered over the past year	Confrontational groups		Non-confrontational groups	
	%	Ranking	%	Ranking
Poor housing conditions, but unable to purchase/build a new home	42.3	3	36.46	3
High child education expenses	25.3	8	20.46	7
Difficulties in disciplining their children	22.0	9	17.13	9
Family disharmony	6.3	12	3.73	12
Heavy medical expenses	34.5	5	30.45	4
Decline in living standards due to inflation	69.0	1	65.23	1
Difficulties in meeting daily expenses due to low household income	51.7	2	44.55	2
Family members' unemployment or job instability	36.3	4	27.35	6
High costs of elderly care	13.8	10	9.47	10
High levels of stress at work	26.3	7	18.51	8
High *renqing* [a] expenditures	33.8	6	28.08	5
Financial losses as a result of scams, theft or robbery	9.2	11	6.31	11
Financial losses from failed investments or business ventures	5.6	13	3.46	13

[a] See note in Chapter I. –Trans.

public satisfaction is an important standard in evaluating the performance of local government and an important benchmark in improving government work (see Table 5.4).

iv. Low levels of government credibility are a motivating factor for confrontational action

Our data show that only small percentages of people would turn to either Party organizations, government departments, labor, youth and women's organizations, or neighborhood committee or village council for help when they encounter problems, Among them about half thought that these entities offered little or no help at all, while another percent thought the same of their employers.

This suggests that more intense confrontational extra-institutional civic engagement by certain social groups is related to the inability of the system to meet their reasonable demands. Under the social structures that prevailed during the initial phase of reforms, civic engagement was mostly carried out through one's *danwei*,[1]

[1] I.e. "work unit". Before China adopted economic reforms in 1978, Chinese society was organized around the *danwei* system. The *danwei* system has since lost most of its relevance, but the term *danwei* remains in popular use as a general term for "workplace". The use of the term here refers to the formal *danwei* system rather than its informal contemporary meaning. –Trans.

Table 5.4 Satisfaction of various social groups with government work (%)

	Previously participated in petitions	Previously participated in protests or demonstrations	Previously participated in labor, school or merchants' strikes	No previous participation in confrontational action
Provision of healthcare services	64.1	72.2	59.8	76.4
Provision of social security	53.7	62.0	55.9	69.5
Compulsory education	79.5	74.0	66.8	87.7
Environmental protection and pollution control	46.9	43.0	37.9	57.2
Crime prevention and public security	60.3	65.8	56.3	71.0
Government integrity and penalties on corruption	27.0	32.0	24.4	40.8
Law-based administration of public affairs	36.1	29.7	28.9	53.3
Economic development that leads to income growth	45.1	57.3	41.1	58.4
Rent regulation and provision of affordable housing for mid- and low-income groups	37.0	53.7	40.8	48.5
Expanding employment opportunities	38.2	58.8	51.8	53.3
Government transparency	28.9	35.7	25.3	48.4

whereas participation by members of the public in social organizations had strong political connotations and undertones of mobilization.

However, social development currently faces issues of misalignment that have arisen during the process of transition from a *danweiren*[2] governance structure to a

[2] Literally "work-unit people". The term *danweiren* implies that citizens are seen as members of work units rather than autonomous individuals. –Trans.

shehuiren[3] governance structure. As China continues to develop its socialist market economy, *danweiren* will play a lesser role in society as the *danwei* system, a remnant of the planned economy era, gradually retreats from its once-dominant role in social governance. However, China has yet to form either an effective social governance structure or a mature system for the constructive engagement and autonomy of *shehuiren*. If *shehuiren* cannot be effectively united to form strong organizational and social identities, society as a whole may very well find itself confronted with various challenges because of inadequate social cohesion.

Meanwhile, the research group also investigated levels of trust among those participating in confrontational action. Their data indicated that levels of social trust remain low in China. In particular, only 51.9% of respondents reported full or strong trust in leading Party/government officials, while only 50% reported full or strong trust in lower-level Party/government officials. To some extent, this reflects the current "crisis of trust" in China. During the course of social transition and re-consolidation of the social order, various institutional arrangements are also being re-adjusted. Some leading Party/government officials do not fully put into action the Party's mass line, are unable to practice the principle of "from the masses, to the masses", and do more work for their practical benefit, which makes it impossible for them to truly receive the support and trust of the public. Hence, it is necessary to establish trust between the government and the people, "listen to the will of the people", "resolve the people's grievances", and create equal rights and opportunities for dialogue between the government and the people.

v. Dissatisfaction with government corruption and local government conduct have become a common motivation for confrontational action

Our data also shows that those participating in confrontational actions have stronger views on government corruption. For instance, 79.9% believe that corruption in China is "very severe" (39.5%) or "rather severe" (40.4%). As for the conduct of local governments, 60% of those participating in confrontational action believe that the government "is concerned only with formalities but not actual work", 58% believe that the government "does not care about the people's welfare", and 58% believe government officials "abuse their authority for personal gain". Meanwhile, those who participate in confrontational action also have a significantly different opinion on local government malpractice. For instance, 32.5% believe that government officials engage in "buying and selling official positions", whereas only 19% do not (see Table 5.5).

Thus, we see that the maintenance of social stability requires the strong addressment of political corruption, stronger institution building, and plugging of loopholes in governance, in order to enhance the government's public credibility. A failure to effectively curb corruption and restore the clean and incorruptible image of the government would lead to social instability and disharmony. Hence, it is necessary

[3] Literally "society people"; *shehuiren*, in the present context, implies that citizens are seen as individual members of society in their own right. The reader may also wish to note that, in other contexts, the term *shehuiren* may carry negative connotations (it is often used as a euphemism for "gangster"). No such connotations are implied in the present context. –Trans.

Table 5.5 Opinions of various social groups on the conduct of local government officials

	% of respondents (confrontational groups)	% of respondents (non-confrontational groups)
Local officials disregard decisions made by the central government	24	17
Local officials are unconcerned about the interests of the people	58	49
Local officials are concerned only with formalities and not actual work	60	51
Local officials engage in misrepresentation and fraud to meet performance targets	50	39
Extravagant spending at the expense of public coffers	55	45
Procrastination and dereliction of duty	54	42
Abuse of power for personal gain	58	45
Buying and selling official positions	32	19

to promote comprehensive political reform, enhance the rule of law and institution building, strengthen the education and oversight of leading officials, and further enhance the Party consciousness of leading cadres.

5.4 A Further Look at the Characteristics of Groups Participating in Civic Engagement

Building upon the previous discussion of our data, we now build a model to further examine the characteristics of groups that participate in civic engagement. Past research has shown that the following propositions generally apply to civic engagement. Firstly, age is usually correlated with the level of civic engagement, where those who are younger are prone to social action. In contrary, older age groups are more conservative in both social consciousness and action. Secondly, civic engagement is highly correlated with educational attainment. Those with a higher level of educational attainment have a higher level of political consciousness and thus a stronger inclination towards civic engagement. Thirdly, groups of higher socioeconomic status are generally more inclined towards political and civic engagement. Because these social groups often command greater social resources, have stronger demands regarding the timely and effective response by politicians to their demands, and have a stronger capacity for social action, they are often known as active participants in political/civic engagement.

Meanwhile, after considering both the theory of quality of life and China's national conditions, I propose the following hypotheses for the purposes of examining the actual state of civic engagement in China.

Firstly, we hypothesize that civic engagement is more likely among those facing greater pressures of life, given that our previous discussion of the data indicates that civic engagement in China is primarily driven by pragmatic interests and social action is motivated by unmet demands, which causes individuals to take actions to protect their rights.

Secondly, we hypothesize that lower levels of social support make it more likely that an individual will participate in social action, given that people are more likely to have their demands met if they have access to greater social resources and wider social support networks. In contrast, a lack of effective social support may motivate an individual to participate in political action in the hope of improving their current circumstances.

Thirdly, we assume that lower levels of trust in the government lead to a greater likelihood of participation in political action. After all, if socio-political action is indeed motivated by unmet demands, it would indicate that such activists have a lower level of trust in government. It is precisely this lack of faith in the ability of government to exercise public powers that motivates them to seek meaningful change through individual action.

Fourthly, we assume that lower levels of social trust lead to a greater likelihood of participation in political action. In other words, if individual political participation is driven by dissatisfaction with society, a conviction that society is unjust would be an intrinsic motivation for political action. Moreover, only individuals who hold such a belief would take proactive action to protect their rights.

These assumptions are based on the national conditions and current state of civic engagement in China. The following general linear model further examines the validity of these assumptions, where we assess the impact of age, educational attainment, social status, burdens of life, social support, and trust in Party/government officials on socio-political participation (see Table 5.6).

The results show that levels of socio-political participation are likely to be higher among males than among females, while urban residents are more active in political participation than rural residents. Age does not strongly influence the level of socio-political participation (coefficient = 0.007), suggesting that all age groups in China may potentially participate in social action. The model also validates our hypothesis that educational attainment is positively correlated with civic engagement. In contrast, the hypothesis that socio-political participation is positively correlated with social status is not supported by the data. In other words, a lower social status predicts a higher likelihood of active socio-political participation.

In addition, the hypotheses proposed in light of national conditions have been verified to be valid. Heavier burdens of life, lower levels of social support, weaker trust in government, and a conviction that society is inequitable all lead to a greater likelihood of participation in socio-political action (significance level of 5%). Our

Table 5.6 Factors underlying socio-political participation

	Coefficient	Standard deviation	*t*-value	Significance level
Constants	0.287	0.127	2.268	0.023
Control variables				
Male (reference group: female)	0.286	0.023	12.626	0.000
Rural residents (reference group: urban residents)	−0.112	0.024	−4.640	0.000
CPC member (reference group: non-CPC members)	0.173	0.040	4.406	0.000
Core variables				
Age	0.007	0.001	7.813	0.000
Educational attainment	0.076	0.007	10.440	0.000
Socioeconomic status	−0.089	0.014	−5.836	0.000
Living costs	0.034	0.005	6.284	0.000
Social support	0.089	0.009	10.191	0.000
Political consciousness	0.128	0.016	8.231	0.000
Trust in leading Party/government officials	−0.048	0.027	−1.783	0.015
Social equity	−0.040	0.020	−1.954	0.051
R	0.321			
R^2	0.101			

Note For the sake of brevity, we have omitted details on how the relevant variables are defined and measured. Specific details can be found in the CGSS 2013

model suggests that civic engagement in China is "defensive" in nature, inspired by a desire to protect one's rights when formal appeals to government authorities are not effective. Participants are chiefly motivated by their low quality of life and dissatisfaction with society. It is hard for civic engagement motivated by such factors to lead to positive engagement between the government and the public; at the same time, such actions are often unconstructive. These issues will be further explored in the following section.

5.5 Civic Engagement and Social Governance

In recent years, social stability and harmony in China have been disrupted by mass incidents, which have arisen in part because China lacks intra-institutional channels for civic engagement. Petitions or other extra-institutional channels are the only way whereby members of the public can express their demands. The *Decision of the CPC Central Committee on Several Important Issues of Comprehensively Deepening Reform*, adopted at the Third Plenary Session of the 18th CPC Central Committee, specifically stated that "the general purpose of deepening [the all-round reform of the socialist market economy] is to develop socialism with Chinese characteristics, to advance modernization in the State governance system and governance capability". It is noteworthy that the *Decision* mentions "social governance", in a departure from the nomenclature favored in previous similar documents, "innovative social management", instead emphasizing "social governance" and innovation that enhances the quality of social governance.

In a healthy society, civic engagement should be based upon constructive interaction with the state. Ideally, the state provides institutional guarantees and lawful channels for civic engagement by establishing and improving systems for civil participation, such as class action lawsuits, public hearings, and lobbying, thus promoting positive interaction between the state and civic organizations. Conversely, in an unhealthy society, a lack of institutional channels for civic engagement may also lead to participation in civic organizations. However, in these circumstances, it would be difficult for civic organizations to interact positively with the state. In contrast, civic organizations may adopt radical modes of action, for instance petitions, demonstrations, and protests, or even "coercion" through mass incidents. In such circumstances, civic organizations may be in opposition rather than interaction with the state. Thus, social development should focus on means for enhancing citizens' trust in the state, and establishing lawful channels for intra-institutional participation by civic organizations.

For a long time, social governance in China has been dominated by the government; China's institutions and structures have offered few channels and opportunities for proactive civic engagement. Meanwhile, civic organizations have neither the resources nor legal latitude to mobilize social forces. This makes it necessary to further strengthen capabilities for civic engagement, enhance channels for civic engagement, and ensure the efficacy of civic engagement at an institutional level. With regards to promoting constructive civic engagement, I propose the following.

First of all, it is necessary to establish and improve channels for civic engagement, and, through political reforms and innovation in social governance, create conditions for civic engagement, expand channels for citizens' direct or indirect participation in social governance and the enactment of social policies, and provide effective guarantees for civic engagement at the institutional level. The experience of Western countries in this respect tells us that, when enacting and implementing policies, broad public participation can greatly reduce government costs and enhance the efficiency of social governance. Thus, the objective laws of social development require that we establish new models of social governance based upon social autonomy and not bureaucracy.

This would require that China, on the one hand, fully protects the rights granted by law to citizens to participate in the political process. On the other hand, it is necessary to expand existing channels of engagement, establish and improve the "stakeholder system", strengthen the role of civic engagement in social governance, and thus transform civic organizations into guardians of public space and public goods. In other words, public participation in social governance will not only be of a non-profit and "fringe" nature. On the contrary, the public may, through civic organizations, exercise sustained autonomy over public spaces and public goods over the long term, and reap the corresponding benefits. These corresponding benefits then constitute effective incentive mechanisms that ensure enthusiasm for and the long-term efficacy of civic engagement, which would maximize the interests of social organizations to the largest possible extent.

Secondly, the existing political structure has long limited civic engagement in China. Because the government has long played a dominant role in social governance through its "top-down" approach, China has yet to develop positive government-citizen interaction. Given the lack of institutionalized channels for their participation, it is difficult for social organizations to form organizational structures that encourage widespread participation by ordinary folk and social elites. In other words, the state's monopoly in all levels of social governance is a direct cause of the barrier between the people and the government.

In another perspective, China has entered an era where social issues are becoming highly prevalent, some of which have an impact far beyond individual interests. Hence, civic engagement oriented towards rights protection commonly manifests as "collective actions". Confrontational mass incidents and potential causes of social instability be nipped in the bud only by improving systems that allow citizens to protect their rights, and making it the norm for citizens to participate in public administration through the establishment of regulated, orderly participation mechanisms. This would reduce information asymmetry between citizens and the government, lead to greater trust in government, encourage citizens to participate in "consultative" proactive civic engagement instead of reactive, "defensive" action taken only when their rights have been infringed, reinforce their "lobbying" capability, and strengthen the standards and institutions governing the "lobbying" activities of social organizations.

In conclusion, a reconstruction of existing social governance mechanisms requires a focus on the transformation of government functions to establish a limited and service-oriented government. Instead of direct intervention in society, the authority of the state should be limited to facilitating the operations of the market economy. At the same time, the state should limit its use of the political administrative system, and promote social integration in place of institutional integration and motivate society's capability for self-governance through the devolution of state powers. This is inevitable if crisis is to be abated and an ideal social governance model be reconstructed. Otherwise, in the face of ever-more complex social interests, the government may find itself overwhelmed by the complexities of daily governance, which would both increase administrative costs and likely lead to crises of public trust as a result of ineffective governance. Specifically, the state should progressively devolve certain functions of public administration (particularly those related to the people's welfare) through political reforms, allow greater participation by the non-governmental sector in public affairs, and accord them corresponding rights in the decision-making process and the right to autonomy. This would then ensure harmony within society even as social rights become more diversified as the space for social autonomy is progressively expanded.

References

Heberer, Thomas. 2005. Zhongguode shehui zhengzhi canyu: yi shequ wei lie [Social and Political Participation in China: The Case of the Neighborhood Communities]. Translated by Lu Lu. *Marxism & Reality* (3).

Huntington, Samuel P., and Jorge I. Dominguez. 1996. Political Development. In *Handbook of Political Science*, ed. Fred I. Greenstein and Nelson W. Polsby. Translated by Chu Fuyun. Beijing: Commercial Press.

Lin, Ka. 2010. Social Quality Theory: A New Perspective for the Studies of Harmonious Society. *Journal of Renmin University of China* (2). (in Chinese).

Nie, Norman, and Sidney Verba. 1996. Political Participation. In *Handbook of Political Science*, ed. Fred I. Greenstein and Nelson W. Polsby. Translated by Chu Fuyun. Beijing: Commercial Press.

Sun, Huan, and Xiaoping Liao. 2010. Ethic Dimension of Subject in Political Participation: Also on Environmental Ethics. *Journal of Beijing Normal University (Social Sciences)* (6). (in Chinese).

Chapter 6
Life Satisfaction and Underlying Factors

Di Zhu

GDP was once the most important indicator for measuring societal development and ensuring human welfare. Most people aspire towards a better life—"money equals happiness" still applies to most situations. Yet, there is evidence that happiness and life satisfaction does not necessarily correspond to material wealth. From 1990 to 2010, even as per capita GDP in China grew at rates exceeding 8%, reported life satisfaction showed a U-shaped trend. By comparing data collected by the World Values Survey, the Asian Barometer Survey, Gallup Inc., the Horizon Research Consultancy, and the Pew Research Center over different periods in time, Easterlin et al. discovered that levels of life satisfaction among Chinese residents began to decline from 1990 onwards. It was not until 2000–2005 that this trend began to reverse. In the face of such contradictions, the international community and academic community gradually came to focus not just on objective material conditions but also subjective assessments of life. This led to broad debate: What significance do economic factors hold in relation to subjective well-being? Does the influence of economic factors vary at the micro and macro levels? Several scholars have also included social equity, social security, social support, eco-environmental factors, health, and marriage in studies on the non-economic factors that influence levels of life satisfaction.

Although there has been a large body of research on life satisfaction and subjective wellbeing, sociologists have yet to make substantial contributions to this field. The influence of social justice and public services, subjects that have been more widely discussed in sociology, on life satisfaction have been emphasized to a lesser degree. Existing research is more concerned with the measurement of life satisfaction, emphasizing the function of uni-dimensional factors such as economic conditions and the environment from an academic perspective. As such, there is lack of comprehensive, systematic observations on the factors underlying life satisfaction. The research undertaken in this chapter was motivated by a desire to rethink the effects

D. Zhu (✉)
Institute of Sociology, Chinese Academy of Social Sciences, Beijing, China

P. Li (ed.), *The Quality of Life in Contemporary China*, Research Series on the Chinese Dream and China's Development Path, https://doi.org/10.1007/978-981-16-7021-3_6

of socio-economic policies and propose policy suggestions. Hence, we propose an analytical framework comprising the factors of economy, society, the environment and individual quality of life, in the hope of providing a response towards relevant policies and institutions. I will also emphasize the role of a social safety net, social justice, and environmental factors in influencing life satisfaction from a sociological perspective. Based on this analytical framework, the present chapter uses data from the 2013 Chinese General Social Survey to build a model to explain life satisfaction, describe the degree of influence of relevant factors and test our analytical framework.

The present chapter first delineates the concept and research significance of life satisfaction before establishing the analytical framework on the basis of existing research. The section on empirical analysis will introduce the sources of our data before analyzing the life satisfaction of urban residents in China and comparing it with other countries and other periods in history. The central part of the section builds our explanatory model for the life satisfaction and discusses research discoveries. The chapter will end by summing up empirical findings and discussing the lessons that they bear for academic research and policy making.

6.1 The Concept and Significance of Life Satisfaction

Unlike objective indicators such as income, education, and health, life satisfaction is a subjective measure of welfare and quality of life, as assessed by people's feelings and experiences. Apart from life satisfaction, other related terms used by scholars writing in English include "subjective wellbeing" and "happiness". "Subjective well-being" is an umbrella term for the different valuations people make regarding their lives, the events happening to them, their bodies and minds, and the circumstances in which they live.

Diener et al. breaks subjective wellbeing into three main components:

1. Life satisfaction, which includes how a respondent evaluates or appraises his or her life taken as a whole, along with individual evaluations of their income, housing, and leisure circumstances.
2. Affect, including positive affect such as joy and satisfaction, and negative affect such as dismay or anger.
3. Domain satisfaction, including sense of achievement and meaning.

"Happiness" is not as well-defined, where there are at least two distinct meanings to this term in an academic context. The first is as a measure of emotional state, for instance: "Were you happy yesterday?" Such emotional experiences are generally short-lived and fluid, and highly influenced by recent events. The second is as a measure of both emotional state and evaluation of life, for instance "On the whole, would you say you lead a happy life?" In this sense, "happiness" is highly similar to "subjective wellbeing". In the World Happiness Report, which consolidates independent reports by researchers from Canada, the UK and the US, the "happiness" of people from various countries are discussed under this framework.

Thus, at a conceptual level, life satisfaction is a major component of subjective well-being. Compared to positive/negative affect and domain satisfaction, which evaluate the emotional or spiritual aspects of life, self-reported levels of life satisfaction which stem from a respondent's assessment of his or her life in terms of employment, income, marriage, housing, and health can more directly reflect socio-economic progress and the effect of social policies. Thus, of the many dimensions inherent in subjective wellbeing, life satisfaction is the one that directly responds to policy measures. This quality is what makes life satisfaction significant to institutional and policy reforms.

Therefore, some policy reports also use life satisfaction as the main means for determining levels of subjective wellbeing, including the OECD How's Life? report. Of course, respondents may not be able to accurately distinguish the difference between life satisfaction and emotional experience during actual surveys. Answers to questions on life satisfaction are often mixed with respondents' emotions and their assessment of various aspects of life. Care must be taken to account for this when interpreting empirical data. At the conceptual level, our chapter makes a distinction between these aspects. Subjective wellbeing is assessed mainly through life satisfaction. Relevant analytical frameworks and empirical concepts will be discussed in later sections.

Indicators that measure life satisfaction have drawn greater attention in recent years, and are now widely used in research on policy and public opinion. Existing studies have pointed out that life satisfaction is a highly sensitive measure of welfare. Despite having a mechanism of action different from GDP indicators, life satisfaction is also an effective measurement tool. As explained by Easterin et al., the GDP indicator represents how material living standards have improved on average, while a life satisfaction index reveals changes in the living standards enjoyed by average citizens, particularly those with lower incomes or levels of educational attainment.

People's material desires have grown along with general living standards, leading to greater concerns regarding issues such as (un)employment, health insurance, children's expenses, and old-age care. As a result, life satisfaction does not increase linearly with rapid economic growth. In this sense, compared with GDP, which is a somewhat simplistic indicator, life satisfaction is more meaningful indicator that comprehensively reflects circumstances of life and wellbeing.

Life satisfaction has also been incorporated in official UK policy. The Green Book published by the UK Treasury (the Exchequer) is an official guide on appraising and evaluating policies. The Green Book makes particular note of how measuring life satisfaction can contribute to Social Cost Benefit Analysis: "At the moment, subjective well-being measurement remains an evolving methodology and existing valuations are not sufficiently accepted as robust enough for direct use in Social Cost-benefit Analysis. The technique is under development, however, and may soon be developed to the point where it can provide a reliable and accepted complement to the market-based approaches outlined above. In the meantime, the technique will be important in ensuring that the full range of impacts of proposed policies are considered, and may provide added information about the relative value of non-market goods compared with each other, if not yet with market goods."

In general, research on life satisfaction includes the three main aspects described below. These aspects are also the main justification for our focus on life satisfaction.

First, life satisfaction is an effective tool for evaluating socio-economic progress and policy systems. Life satisfaction is founded upon subjective experience. As such, it can measure the subjective impact of socio-economic changes on people's lives, thus providing a basis for appraising social progress and the effect of social policies. Despite sustained economic growth, people's feelings about the various aspects of their lives may very well head in the opposite direction. Therefore, effective institutions and policies should not only ensure that citizens are able to meet their material needs, but also seek to enhance their life satisfaction.

Second, research on life satisfaction makes it more feasible to design cross-cultural and transnational socio-economic policies. It is near impossible to design such policies simply by controlling for objective variables, while life satisfaction is quite sensitive to many economic, social, political, and environmental indicators, allowing it to provide information on a wider range of policy consequences than objective indicators. For instance, what are the advantages and disadvantages of the same policy in different national and cultural contexts? Of course, there still certain conditions that limit the effectiveness of such comparison.

Third, an analysis of the factors underlying life satisfaction is a robust empirical tool for testing socio-economic outcomes. Such analysis can reveal which factors play a stronger role in enhancing subjective wellbeing, thus allowing us to verify whether outcomes generally regarded as signs of socio-economic progress truly meet people's life expectations and enhance their quality of life. Given that life satisfaction is an important part of overall social welfare, such analysis provides robust empirical evidence for testing the effects of social policies.

However, it should be emphasized that, despite being an important tool for assessing the quality of life, life satisfaction cannot replace objective indicators such as GDP, income, and employment. Some people may be happy despite being poor, but their subjective emotional experience does not translate to a higher quality of life. Neither can governments evade their duties to eliminate poverty and improve livelihoods. Thus, relevant reports and studies on life satisfaction have emphasized that subjective wellbeing is a complement to objective indicators. Much as objective socio-economic indicators do not represent social welfare, subjective wellbeing does not in itself represent quality of life.

6.2 An Analytical Framework for Life Satisfaction

There has been much research on life satisfaction and subjective wellbeing, both inside and outside of China, mostly in the fields of economics or psychology. Most existing research can be divided into two categories. Psychologists, and some sociologists, focus on the measurement and comparison of life satisfaction. Significant research in this respect include Diener et al., Xing (2008), Zhou and Ren (2006), Feng and Yi (2000). Other studies focus on the factors underlying life satisfaction. For instance, such research may focus on "whether money equals happiness", i.e. the effect of income or GDP on life satisfaction. In this regard, Easterlin made an observation that would come to be known as the Easterlin paradox, where he found that greater household income is correlated with greater life satisfaction, but average levels of life satisfaction for a country do not appear to increase over time in line with increases in average national income. There is still debate among scholars regarding how income levels affect life satisfaction, which we will address in greater detail later.

There are also studies that focus on environmental, psychological, or employment factors (Cao 2011). Yet other studies have focused on the composite effect of multiple factors. For example, Fang and Feng (2009) analyze subjective wellbeing from the perspectives of employment, income, housing, health, and interpersonal relationships. Liu et al. (2012) introduced annual income, subjective ratings of social status, changes in social status, political identity and other such demographic variables in their analysis of the factors underlying levels of happiness. The urban life satisfaction model constructed by Appleton and Song included income, health, health insurance, political participation as explanatory variables along with related control variables.

Existing studies have measured life satisfaction, happiness, and subjective quality of life, and their relationship with income in great detail. However, comprehensive analysis of the factors underlying life satisfaction has been lacking. Alternatively, we can say that such research has paid less attention to the development of frameworks for analyzing such factors. Why are certain underlying factors selected? What are the relevant underlying theoretical assumptions and theoretical concerns? What are the internal structures of these factors and how are they connected? Only by elucidating these issues can we ensure the reliability and validity of an explanatory model.

More countries and social organizations now realize that improvements to citizens' quality of life and wellbeing require due consideration of economic, social and environmental factors. This has weaned them of their sole reliance on GDP as an indicator in measuring social progress. More comprehensive indicators are now employed, the more well-known of which include the Human Development Index,[1] which emphasizes health and education, the Happy Planet Index,[2] which emphasizes sustainable, eco-friendly development, and the Better Life Index,

[1] United Nations Development Programme, http://hdr.undp.org/en/humandev.

[2] The New Economics Foundation, http://www.happyplanetindex.org.

which comprehensively evaluates material wealth, quality of life, and sustainable development.

There is much existing research on the relationship between life satisfaction and such factors as income, social security, social equity, quality of public services, the environmental quality, and individual quality of life, which provide an approach to designing our analytical framework.

6.2.1 Life Satisfaction and Economic Factors

With respect to economic factors, existing research has mainly focused on the effect of income on life satisfaction. In addition, by making a distinction between absolute and relative perspectives, researchers have emphasized that there are different mechanisms by which life satisfaction is influenced by income. Based on a demand-satisfaction framework, Veenhoven postulated that happiness is preconditioned upon the satisfaction of innate basic needs. Because those with higher incomes are able to satisfy their basic needs, their levels of income and happiness are positively correlated, and their happiness is not "relative", i.e. not derived from comparison with others. Of course, Veenhoven also pointed out that human needs do not increase indefinitely, meaning that the relationship between income and happiness obeys the law of diminishing marginal returns. Past a certain threshold, the marginal utility of income diminishes to zero.

Unlike Veenhoven, Easterlin argues that happiness is relative. In other words, satisfaction is "relative", derived from comparison with one's previous status or from comparison with others. Thus, Easterlin believed that income levels do not induce a change in life satisfaction. Rather, it is material desire, as an intermediary variable, that affects life satisfaction: an increase in income brings about greater material desires, while greater material desires offset any positive effect by income growth on life satisfaction. The "relative" view is also held by Diener and Biswas-Diener, whose review of relevant literature found that: (1) income growth raises life satisfaction only for those in extreme poverty, not for those of the middle or upper-middle class, or residents of rich countries; (2) material desires and the drive to accumulate greater wealth are factors that thwart any increase in life satisfaction.

Chinese scholars have derived relevant empirical evidence from domestic data that shed further light on these debated issues. Xing (2011) noted the relationship between absolute income and happiness. By analyzing data from surveys carried out in six Chinese provincial capitals, Xing found that income had only a slight effect on happiness for urban residents, while there was a stronger correlation between income and happiness in less affluent regions. Xing also found that people with higher incomes were generally less happy than those with lower incomes. Luo (2009) defined absolute income as total household income per capita, while relative income was defined with reference to the subjective poverty line, average income in the locality, and previous income levels. After analyzing data from randomized surveys of urban and rural residents across China, Luo found that relative income has a significant impact on subjective wellbeing. In addition, even after controlling for relative income, absolute income also had a significant impact on the same.

Guan (2010) used mathematical techniques to treat absolute income as income for the past year and relative income as individual assessments of one's income level in an analysis of survey data from Beijing, Shanghai, and Guangdong. Guan's results show that relative income, but not absolute income, had a significant impact on happiness. Guan also used regional characteristics, educational attainment, age factors, and the population quantile method to construct four different indicators to measure relative income and build corresponding regression models, all of which verified that relative income indeed had a significant impact on happiness.

Using a comparative framework, Liu et al. (2012) included relative social status (relative to other persons) and improvements in social status (compared to past status) as explanatory variables for levels of happiness. Their model showed that relative social status, both in comparison with others and one's past status, had a significant impact on happiness, suggesting that "the pursuit and achievement of a higher relative status is an important factor underlying levels of happiness".

6.2.2 Life Satisfaction and Social Factors

Existing studies mainly discuss the impact of social factors on life satisfaction from the perspectives of social security, social governance, social equity, public services and social support. Diener and Suh argued that, in addition to economic indicators, social indicators and subjective wellbeing should be included in measures of "the good life", i.e. wellbeing and quality of life. Such social indicators include health indicators (infant mortality rates, number of doctors, life expectancy), public safety and security (suicide rates, size of the police force, rate of sexual violence), as well as indicators related to the environment, human rights, social welfare, and education. These social indicators also have a large impact on subjective wellbeing.

Diener and Sellman emphasize the role of non-economic indicators in measuring national wellbeing, including social governance (freedom, democracy, political stability), social capital (trust, civic engagement), and religious beliefs. Dolan et al. pointed out that income inequality has a negative impact on life satisfaction, but this effect differs across different countries and groups. For instance, Latin Americans are happy despite greater income inequality, while people from the former Soviet bloc are less so despite more equitable income distributions. Needless to say, whether or not income inequality adversely impacts life satisfaction depends on people's views: does income inequality benefit economic efficiency, or does it lead to unequal opportunities? Researchers have also found that social mobility is lower in Europe, meaning that income inequality leads to unequal opportunities, which needless to say has a negative impact on life satisfaction.

When analyzing macro data, scholars generally tend towards using objective indicators such as the Gini coefficient to measure the impact of social (in)equity on life satisfaction. When using micro data, however, it is more common to measure such impact on life satisfaction through attitudes towards social (in)equity. For instance, Shi and Cui (2006) used data from nationwide surveys to explore the relationship between social equity and life satisfaction. Their study defined citizens' views of social equity from two aspects: (1) attitudes towards just outcomes, procedural fairness, and social regulation; (2) appraisals of one's current social standing. Attitudes towards just outcomes refer to attitudes pertaining to modes of income distribution and economic inequality, attitudes towards procedural fairness refer to attitudes pertaining to fairness of opportunity and the market economy, while attitudes towards social regulation reflects whether one's values are "individualistic" or "group-oriented".[3] Their research concluded that life satisfaction was significantly influenced by procedural fairness and appraisals of one's social standing, but not just outcomes.

Few studies have explored the impact of social security on life satisfaction. Di Tella et al. used pension replacement rates to approximate the level of social security. By analyzing data on individual pensioners in Europe, they found that a higher pension replacement rate enhanced life satisfaction for both the employed and unemployed. Appleton and Song analyzed data from surveys conducted in urban Chinese regions and found that health insurance had a significant impact on life satisfaction. Those with coverage for major health conditions (part of the premiums is borne by the individual) and those without any form of health insurance had lower levels of life satisfaction than those who were entitled to health insurance provided by the state.

6.2.3 *Life Satisfaction and Environmental Factors*

In related research, there are usually two types of methods for measuring environmental factors. The first type uses objective indicators such as levels of nitrogen dioxide and suspended particulate matter, while the second type uses subjective indicators such as environmental awareness and environmental satisfaction. Most empirical research suggest that levels of environmental pollution and environmental concerns are significantly correlated to life satisfaction. This emphasizes the role of sustainable development and consumption in enhancing life satisfaction.

Makeron and Mourato collected data on air quality in 400 London homes, and asked their inhabitants to indicate whether they felt air pollution was a serious issue in their neighborhood. In so doing, they discovered that both objective measurements and subjective appraisals of air pollution levels had a significant adverse impact on life satisfaction. This was true even after controlling for other related variables. Smyth et al. focused on the relationship between air pollution and subjective wellbeing in Chinese cities. Their study assessed environmental quality through levels of sulfur dioxide emissions, rates of environmental incidents, traffic congestion levels, and

[3] In other words, whether one values the group over oneself or vice versa. –Trans.

availability of green spaces, while controlling for socio-economic variables and political views, such as income inequality, opinions on the merits of the market economy, social protections, as well as respondents' personal traits. They found that severe air pollution, greater rates of environmental incidents, and severe traffic congestion significantly lowers subjective wellbeing. Approval of the market economy system and a belief that society is unequal also results in lower subjective wellbeing. In a study that involved 18 cities, Cao (2011) used nitrogen dioxide concentration levels to measure air quality and found that there is a significant correlation between air quality and life satisfaction among low-income groups. The same was not true for those with higher incomes.

Environmental consciousness is also included in subjective indicators. After a panel survey of British households that controlled for environmental conditions, residential localities, lifestyles, and psychological characteristics, Ferrer-I-Carbonella and Gowdyb found that concern about ozone pollution levels was negatively correlated with life satisfaction. They also found that life satisfaction was more affected by the degree of concern for the environment than objective levels of environmental pollution. In addition, concern about biodiversity was also positively correlated with levels of life satisfaction. The authors interpreted this to mean that concern for the surrounding environment brought about positive psychological affect.

Since the environment has a major impact on human wellbeing and life satisfaction, both researchers and the international community advocate a balance between "environmental consumption" (development and utilization of environmental resources) and human wellbeing. The pursuit of human wellbeing may well be the fundamental goal of social development, but lower environmental impacts and more efficient utilization of environmental resources is key to sustained improvement of human wellbeing. This idea has become known as the "environmental efficiency of wellbeing (EWEB).

Knight and Rosa used variables that affected life satisfaction in a regression model to measure EWEB. Using data from 105 countries, they examined how EWEB was related to economic development, democracy, inequality, levels of trust, and climate factors.

Interestingly, economic development was related to EWEB by a negative quadratic function, implying that EWEB increases when the level of economic development is low, but declines when an economy becomes more developed. In turn, this implies that life satisfaction improves only through greater environmental consumption. This refutes the theory that only economic growth can sustain greater human wellbeing. In addition, social inequality is negatively correlated with EWEB while levels of trust are positively correlated. This "sustainable wellbeing" framework, which accounts for environmental impact, bears important lessons for our analytical framework and relevant policy proposals.

Jacob et al. explored the impact of ecologically sustainable behavior on subjective wellbeing from the perspective of sustainability. In particular, they examined the impact of sustainable food consumption, one of the indicators of ecologically sustainable behavior. Sustainable food consumption generally refers purchasing locally grown or organic food, dining in restaurants operated by local residents, and cooking at home. Their empirical analysis showed that sustainable food consumption had a significantly positive impact on subjective wellbeing. The authors interpreted this to mean that reflection and self-discipline leads people to put greater store upon inner experiences and processes, which then leads to greater satisfaction and happiness.

6.2.4 Life Satisfaction and Individual Quality of Life

When measuring life satisfaction, existing research also emphasizes the impact of health, time and unemployment on an individual's quality of life. Among the three, health and unemployment have been found to have a greater impact. For example, in an analysis of papers published in economics journals since 1990, Dolan et al. found that despite a lack of consensus, or even disagreement, among researchers on the factors underlying subjective wellbeing, scholars generally agree that health and unemployment are significant factors. This has been supported by evidence from different countries, different time periods, and different datasets, compiled using a variety of research methodologies.

Through an analysis of data from EU countries, Segheri et al. showed that higher self-reported levels of health are highly correlated with greater life satisfaction. Appleton and Song found that this was also true for urban residents in China. Of those who believed they were "very unhealthy", only 14% were satisfied or very satisfied with their lives. In contrast, of those who believed they were very healthy, 44% were (very) satisfied. The authors interpreted this to mean that ill health brought about negative emotions and financial burdens, which in turn negatively impacts life satisfaction.

Appleton and Song also pointed out that unemployment has a strong impact on the life satisfaction of urban residents in China. Unemployment has a significant direct impact on life satisfaction, even after controlling for income and other factors. Diener and Biswas-Diener found that most existing studies have pointed out that unemployed persons are significantly less happy, even after controlling for income. Their paper attributed this to cultural norms, as a lack of unemployment leads to feelings of inadequacy. The paper also noted that unemployed persons often receive less respect, especially for men. However, Chinese scholars often neglect to account for the social pressures arising from unemployment when constructing analytical models of life satisfaction. Alternatively, they take a one-dimensional view of unemployment (i.e. whether one has a job), without considering the considerable range of factors involved. Obviously, unemployment due to *xiagang*[4] is different from unemployment due to personal reasons. A failure to account for such factors weakens the causal relationship between unemployment and life satisfaction. For instance, the model built by Guan (2010) found that life satisfaction is not significantly impacted

[4] See note in Chapter II. –Trans.

by whether one has a job. This serves as a reminder to exercise care in selecting indicators when analyzing the impact of unemployment and accompanying pressures on life satisfaction.

As for the time effect, a literature review by Dolan et al. found how one spends their time strongly influences life satisfaction. Existing studies have focused on how subjective wellbeing is affected by the frequency and time spent on work, commuting, caring for others, volunteer work, exercise, and religious activities.

6.2.5 Establishing our Analytical Framework

Based upon existing research on subjective wellbeing, happiness, and life satisfaction, this chapter argues that life satisfaction is more appropriately analyzed through a framework that includes economic, social and environmental factors. At a practical level, these factors provide more comprehensive insights into life satisfaction, while also providing a more reliable theoretical and empirical basis for policy formulation and implementation. At a technical level, a more comprehensive analytical framework reduces errors and allows us to more accurately explain the impact of the various relevant factors.

Firstly, one's economic standing, as measured mainly through income, provides the material conditions necessary for sustaining daily life. Although there is still much debate among researchers regarding the impact of income on life satisfaction, there can be no doubt that income cannot be excluded as an underlying factor. Secondly, one of the goals and motivations for our chapter is to re-examine institutions and policies that promote socio-economic development. Therefore, in addition to economic factors, we also consider social and environmental factors related to such policies and institutions. Relevant social factors mainly refer to relevant social systems and public services, as well as the evaluation thereof, while environmental factors mainly refer to environmental conditions in the locality of residence. Existing research has shown that, of the social factors, levels of social security, social equity, and quality of public services have a significant impact on life satisfaction. Meanwhile, significant environmental factors include objective environmental indicators and subjective opinions on environmental quality. In addition to factors of the external environment (i.e. the social environment and the natural environment), an individual's life satisfaction is also related to their lifestyle and quality of life. Current research has clearly indicated that an individual's life satisfaction is also related to their health, employment status, and how the individual allocates their time across various activities. The analytical framework for this chapter is illustrated in Fig. 6.1.

Our analytical framework includes economic factors (the material conditions for sustaining daily life), non-economic factors (social systems and public services, eco-environmental quality, and individual quality of life), the external environment (social factors and environmental factors), and internal factors (material conditions and individual quality of life). These factors are largely mutually independent. Figure 6.2 shows the structure of our analytical framework. In particular, the level of economic

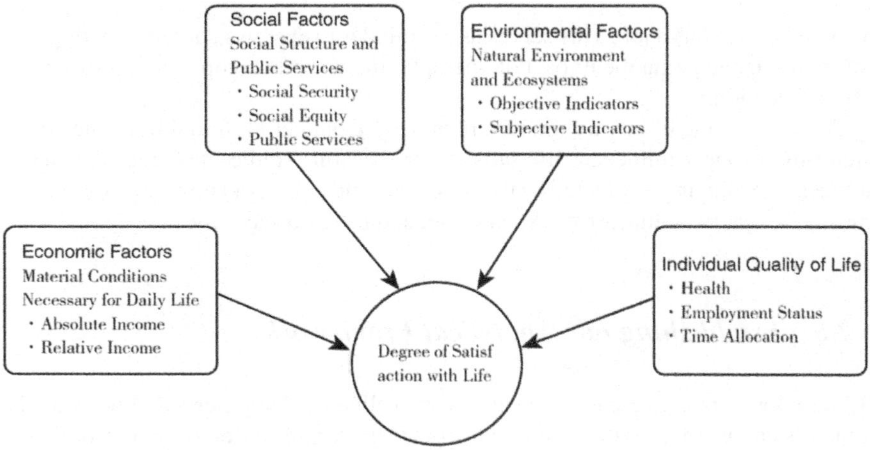

Fig. 6.1 Analytical framework for the factors underlying life satisfaction

development, an economic aspect of the external environment, is a macro-level variable. As our chapter uses micro data, such factors have not been included in our analysis.

This analytical framework constructed from the abovementioned theoretical assumptions is able to more comprehensively account for differences in life satisfaction and thereby improve the reliability of our analysis. Although our chapter has argued for the use of comprehensive analytical frameworks, it remains impossible to include all possible underlying factors. Instead, we have opted to only include those factors important to our purposes and relevant to our theoretical assumptions

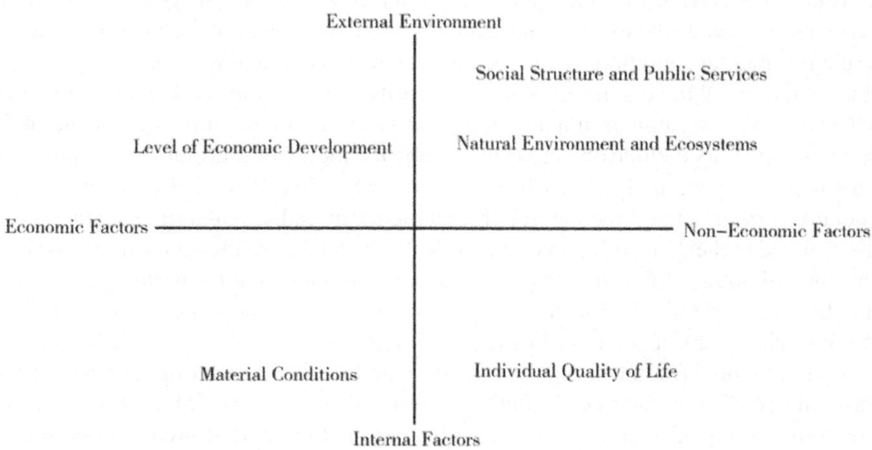

Fig. 6.2 Structure of the life satisfaction analytical framework

in our framework, the validity and robustness of which requires further empirical verification.

6.3 Comparing Life Satisfaction Levels

Our chapter uses data from the 2013 Chinese General Social Survey, coordinated by the Institute of Sociology at the Chinese Academy of Social Sciences. The survey uses a multi-stage random sampling method that covers 30 urban/rural areas in 30 provinces/autonomous regions/municipalities directly under the central government. The sampling design ensures that the data is respectively representative of urban and rural areas. The CGSS 2013 collected 10,206 valid responses from Chinese citizens aged 18 or above.

In China, the factors underlying life satisfaction vary widely between urban and rural localities, given that the social safety net and quality of public services are poorer in rural areas. This means that, although both social security and public services significantly impact life satisfaction in both urban and rural localities, there may be differences in the underlying mechanisms. For this reason, our chapter will focus on the life satisfaction of urban residents.

The CGSS 2013 includes a set of questions that measure life satisfaction. Respondents are first asked to rate their satisfaction with their educational attainment, health, social life, family relations, household economic situation, leisure, and entertainment, before being asked to rate their overall satisfaction ("On the whole, how satisfied are you with life?") Respondents are asked to rate their answers on a scale of 1–10. A score of 1 means "very dissatisfied" while a score of 10 means "very satisfied". Average score for urban residents was 6.84.

Using data from the Gallup World Poll, the World Happiness Report reports average happiness levels in China from 2010 to 2012 to be 4.978 (out of 10), an increase of 0.257 over levels from 2005 to 2007. This put China in 93rd place among 156 countries. The World Values Survey revealed a similar trend, where life satisfaction was 6.85 (out of 10) for the average Chinese resident, putting China in 31st place among 58 countries, a slight improvement over 2005–2009 rankings, where China was in 36th place.

As previously discussed, analysis by Easterlin et al. found that life satisfaction in China from 1990 to 2012 followed a U-shaped curve. As shown in Fig. 6.3, domestic data indicates that social equity was higher in China in 1990, when the average Chinese resident scored their life satisfaction at 7.29. Life satisfaction dipped to a low of 6.53 in 2001 before rebounding somewhat. Liu Junqiang et al. (2012) used CGSS data to stress that happiness levels rose steadily from 2003 to 2010 and argued against the Easterlin paradox. Based on a five-point scale, the level of happiness was 3.27 in 2003 and 3.77 in 2010. If respondents who indicated that they are "happy" and "very happy" are categorized as one group (similarly for those who are "unhappy" or "very unhappy"), 72.6% of respondents were happy with life in 2010, up from 37.3% in 2003, while only 9.8% were unhappy in 2010, down from 12.9% in 2003.

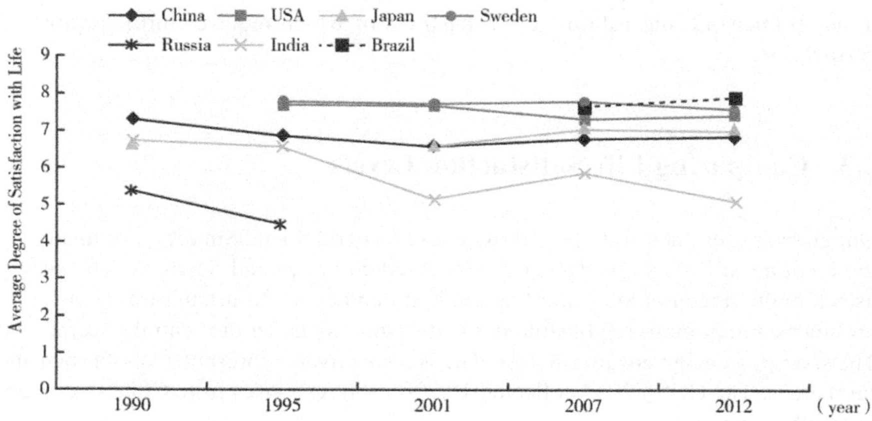

Fig. 6.3 Life Satisfaction in China, BRICS nations and Developed Countries (World Values Survey). *Source* http://www.worldvaluessurvey.org/WVSOnline.jsp. The horizontal axis represents the years that China conducted relevant surveys. Corresponding surveys for other countries were conducted within two years of the corresponding year in the horizontal axis

We have also compared the life satisfaction of Chinese residents with BRICS countries, major developed countries, and Nordic countries with strong public welfare systems. As shown in Fig. 6.3, Chinese residents reported lower levels of life satisfaction than residents of the US (representative of developed countries) and Sweden (representative of countries with a strong public welfare system). Life satisfaction was also lower in China than in Japan post 2001. However, as far as trends are concerned, US citizens have reported lower life satisfaction levels since 2001, while the reverse is true for China. The level of life satisfaction in Brazil is comparable to that of the US and Sweden, and also much higher than that of other BRICS nations. Life satisfaction is relatively low in India and Russia.

A comparison with other countries not only reveals differences in the level of economic development, but also differences of culture and history. Although the US has been affected by such factors as the economic crisis and a widening wealth gap, the overall strength of its economy is still able to sustain a high standard of living for its citizens. Thus, life satisfaction levels in the US remain relatively high, on par with developed countries with strong public welfare systems. Brazil's relatively high level of life satisfaction is largely related to its unique cultural and social context. Residents of Latin American countries are more oriented towards desirable goals and positive emotions. Meanwhile, lower life satisfaction levels in Russia are probably related to social instability, lower incomes and ideological crisis after the dissolution of the Soviet Union. However, data indicates that life satisfaction in Russia has improved greatly since 2007.

Easterlin et al. attributed changes in life satisfaction in China to greater materialistic desires, while Liu et al. (2012) emphasized that economic development leads to higher levels of happiness. In reality, there is no contradiction between the trends unearthed by these two studies, as well as the attributed underlying factors. In the 1990s, market reforms were necessary for restoring China's economy and raising living standards, but a widening rich-poor divide and upheaval of existing moral values eventually resulted in a sustained decline in life satisfaction and happiness. Since 2003, both the government and the general public have begun to rethink the consequences of "blindly pursuing GDP growth". A renewed focus on income equity, livelihood security and cultural development led to a rebound in life satisfaction levels. Materialistic desires stimulated by economic growth may also be interpreted as a response to less-than-ideal institutions and policies, which has created new imperatives for reform. Therefore, despite the "negative" impact of greater materialistic desires, we should also observe that new priorities that lead to more comprehensive socio-economic development, during the initial and "maturity" phase of market reforms, may just be key to understanding trends of change in life satisfaction.

Figure 6.4 shows the life satisfaction of Chinese citizens, as derived from data from the CGSS 2013. Satisfaction with family relations (8.26), social life (6.99), and health (6.95) are greater than overall life satisfaction, indicating that these factors play a role in enhancing overall life satisfaction. Meanwhile, overall life satisfaction is lowered by satisfaction with educational attainment (5.66), household economic situation (5.81), and leisure and entertainment (5.83).

In our analysis, overall life satisfaction is the dependent variable while satisfaction with various aspects of life serve as independent variables. We then construct a regression model to examine how, and to what degree, satisfaction with the various aspects of life affect overall life satisfaction. As shown in Table 6.1, our overall model is statistically significant and can account for 56.2% of the variability in overall life satisfaction.

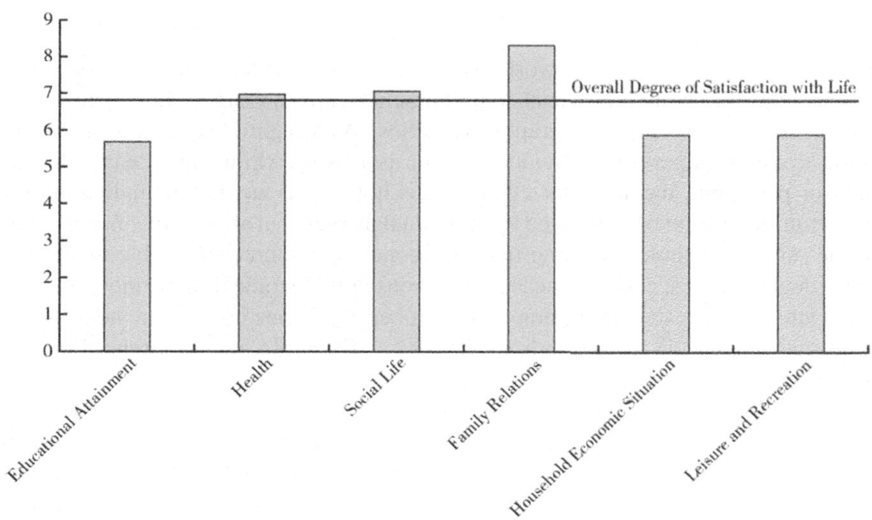

Fig. 6.4 Life satisfaction of urban residents (by category) (1 = very dissatisfied, 10 = very satisfied)

Table 6.1 Decomposition model for overall life satisfaction (linear regression model)

Satisfaction with various aspects of life	Coefficient	Satisfaction with various aspects of life	Coefficient
Educational attainment	0.070 *** (0.008)	Household economic situation	0.291 *** (0.010)
Health	0.064 *** (0.009)	Entertainment and leisure	0.222 *** (0.009)
Social life	0.067 *** (0.011)	Constant	0.717 *** (0.092)
Family relations	0.222 *** (0.011)	Sample size	5539
		Adjusted R^2	0.562

Note Numbers in brackets represent robust standard errors for the relevant coefficients
*** $p < 0.01$

Individual variables of our model are also statistically significant. After controlling for other variables, satisfaction with one's household economic situation has the greatest impact on overall life satisfaction. This indicates that, in our measurement framework, superior material conditions exert the largest impact on increased life satisfaction for urban residents. Meanwhile, satisfaction with family relations and leisure/entertainment are the second-most important factors underlying greater life satisfaction. In contrast, levels of satisfaction with social life, health, and educational attainment have a lesser impact.

6.4 Analysis of the Factors Underlying Life Satisfaction

Based on our analytical framework, we will seek to elucidate how life satisfaction is impacted by economic, social, environmental, and quality-of-life factors after controlling for relevant demographic variables. With regards to economic factors, some academic papers use individual income as a proxy, while others refer to household or per capita income. Our chapter uses household income, as individual life opportunities are not only affected by individual income, but also the income of others in the same household, meaning that the economic resources available to any one individual is more accurately measured by household income. Furthermore, respondents adopt inconsistent definitions for "number of persons in the same household", resulting in larger disparities in household size. Therefore, total household income is used to measure a respondent's absolute income.

As for indicators that measure relative income, a lack of tracking data makes it impossible to determine changes in household income. However, the survey asked respondents to rate how their socio-economic status and living standards have changed over the past five years. This allows us to approximate both vertical changes (i.e. current income compared to past income) and horizontal changes (i.e. income levels relative to one's peers). Two methods are commonly used in existing research to measure relative income levels (relative to one's peers). The first is the "objective measurement method". For instance, this method would measure the gap between an individual's income and the average income of persons of similar educational attainment, age, or income interval. The other method is the "subjective evaluation method", which asks a respondent to select a subjective benchmark and rate their income level accordingly.

Guan (2010) advocates the subjective evaluation method, arguing that no system of objective indicators can accurately reflect a respondent's subjective appraisal of their income level. In Guan's study, respondents are commonly asked to rate "their socio-economic status at the local level". Although "local level" may be overly vague, the phrasing of this question allows respondents to indicate the (subjective) geographic scope of their social relations and relevant benchmarks, making it more relevant as a "horizontal" measure of relative income. Guan (2010) also used relative income (defined using various reference indicators) to build relevant models that demonstrate the robustness of using an individual's subjective appraisal of their income status to predict their level of life satisfaction.

As for social factors, our chapter focuses on three aspects: social security, social equity and public services. In existing research, levels of social security are often measured by health and old-age insurance. As old-age insurance for Chinese civil servants and employees of public institutions have not been fully integrated, they do not contribute towards old-age insurance funds. In addition, our data does not include detailed variables for persons not entitled to old-age insurance coverage. These factors make it impossible to measure the level of social security through objective indicators related to old-age insurance. In addition, a large number of respondents enjoy health insurance coverage. However, the specific type of health insurance is not related to their level of social security, meaning that objective health insurance indicators are not reliable indicators of social security. In contrast, subjective indicators, i.e. satisfaction with old-age and health insurance, are more effective in measuring the level of social security. In questions related to social equity, respondents are most concerned with the rich-poor divide and income distribution. Thus, we assess the level of social equity through respondents' ratings of "wealth and income distribution" equity. Public services are often related to the necessities of daily life. Our chapter uses respondents' ratings of government efforts regarding healthcare and housing security as a proxy for the quality of public services.

The satisfaction of respondents with the environmental quality in their locality of residence is used as a proxy for environmental factors. As for individual quality of life, the present chapter mainly focuses on health, employment status, and working hours. Although our data lacks variables that directly measure health status, there are a group of variables that measure the burdens of everyday life, including "heavy medical expenses", which reveal the mechanisms by which health status influences life satisfaction. Because these provide a more accurate means for measuring the impact of health on life satisfaction, the present chapter opts to evaluate health factors using such variables that measure economic burdens. Many researchers have also noted that unemployment is a factor that strongly affects life satisfaction. In the CGSS 2013, 3.6% of urban respondents are currently unemployed, albeit for various reasons, including employer-related factors (bankruptcy, enterprise reform, *xiagang/neitui/maiduan gongling*,[5] forced resignation), personal factors (family reasons, health, resignation), failure to find a job after graduation, and family responsibilities. Thus, a respondent's life satisfaction level cannot be effectively explained by his or her employment status. Additionally, there are technical difficulties in identifying whether a respondent is "truly unemployed".

Similar to how health factors are measured, the present chapter measures unemployment risks faced by a respondent or their family members by asking if respondents face psychological pressures due to the unemployment or job instability of a family member. This clearly defined "pressure—satisfaction" theoretical assumption allows us to more accurately measure the impact of unemployment. In the CGSS 2013, there is only one variable regarding working hours—"Since the start of the year, how many hours have you spent per day in your nonfarm job?" As more than 1/3 of respondents are currently out of work, while 12% of employed respondents are full-time farmers, we have had to omit this variable, given that its inclusion would result in the loss of much respondent data.

A range of demographic variables commonly used in existing research were included as control variables, including gender, age, marital status, household registration, and education attainment. Occupational factors were omitted despite their relevance, given their significant correlation with educational attainment (a correlation coefficient of 0.3324). Only 5% of sample data originate from Beijing and Shanghai (Guangzhou and Shenzhen were not included in the survey), while 47% of respondents did not indicate whether they were Communist Party members. Because incomplete information affects the reliability of statistical inference, residential locality and Party membership were omitted from our analysis, despite being commonly used demographic variables. Variables included in our model, along with descriptive statistics, are listed in Table 6.2. Our model applies the natural log transformation on total household income to reduce collinearity and heteroscedasticity.

Overall life satisfaction, measured as a continuous variable, is the dependent variable in our stepwise regression model. Economic, social, environmental and individual quality of life factors are successively introduced as explanatory variables to test the ability of our model to account for changes in the independent variables. Appropriate weights are also given to age and gender to make our samples more consistent with population demographics (based on China's sixth census). Our various linear regression models are described in Table 6.3.

Our regression results show that all models are able to explain changes in life satisfaction, from Model 1, whose independent variables include only absolute income, to Model 6, which includes all relevant variables. The explanatory power of our model increases as successive independent variables are introduced.

Model 1, where total household income is used as a proxy to measure absolute income levels, is able to explain 4.4% of variation in life satisfaction. Total household income is positively correlated with life satisfaction at a statistically significant level.

---

[5] See note in Chapter II. –Trans.

Table 6.2 Descriptive statistics for relevant dependent and independent variables

Variable	Value/definition	Sample size	Mean	Standard deviation
Overall life satisfaction	1 = very dissatisfied; 10 = very satisfied	5574	6.842	1.819
Household income (past year)	CNY 0–1.0100000	5191	74,383.92	177,796.3
Household income (past year) (ln)	6.492–16.130	5170	10.806	0.890
Change in living standards compared to five years ago	1 = large decline; 2 = some decline; 3 = no change; 4 = some improvement; 5 = large improvement	5542	3.726	0.950
Relative socio-economic status in locality of residence (subjective rating)	1 = low; 2 = below average; 3 = average; 4 = above average; 5 = high	5490	2.357	0.893
Satisfaction with old-age insurance	1 = very dissatisfied; 10 = very satisfied	5009	6.276	2.488
Satisfaction with health insurance	1 = very dissatisfied; 10 = very satisfied	5278	6.553	2.371
Wealth and income distribution	1 = very unfair; 2 = unfair; 3 = fair; 4 = very fair	5171	2.149	0.732
Appraisal of government rent regulation and affordable housing policies	1 = very poor; 2 = poor; 3 = good; 4 = very good	4467	2.406	0.786
Appraisal of government-provided healthcare services	1 = very poor; 2 = poor; 3 = good; 4 = very good	5304	2.767	0.633
Satisfaction with environmental quality in locality of residence	1 = very dissatisfied; 10 = very satisfied	5582	6.250	2.009
Respondent faces heavy financial burdens due to high medical expenses	1 = yes; 0 = no	5549	0.278	0.448
Respondent faces pressure due to a family member's unemployment or job instability	1 = yes; 0 = no	5556	0.289	0.453

(continued)

Table 6.2 (continued)

Variable	Value/definition	Sample size	Mean	Standard deviation
Gender	1 = male; 2 = female	5583	1.542	0.498
Age	1 = 18–25 2 = 26–35 3 = 36–45 4 = 46–60 5 = over 60	5583	3.153	1.208
Marriage status	1 = married (including cohabitating couples and first/subsequent marriages); 0 = single (never married, divorced, or widowed)	5575	0.816	0.387
Educational attainment	0 = primary education & below; 1 = junior secondary; 2 = upper secondary (high school/secondary vocational education); 3 = higher vocational education; 4 = bachelor's degree and above	5575	1.553	1.261
Household registration	1 = rural; 2 = urban	5569	1.536	0.499

Estimates in Model 1 are based on adjusted R^2 values. Changes in the explanatory power of the model are examined by successive introduction of other independent variables.

We first introduce variables related to relative income levels—current income versus past income and individual income versus income of one's peers to form Model 2. Absolute income serves as a proxy for living standards, while relative income serves as a proxy for self-assessed socio-economic status. In Model 2, we find that all three variables are positively correlated with increased life satisfaction at a statistically significant level. Model 2 is able to explain 17% of variation in life satisfaction, which clearly indicates that relative income has a large impact on the life satisfaction of urban residents.

In Model 3, we introduce five variables to measure three different social factors: level of social security, level of social equity, and quality of public services. We observe another significant rise in the explanatory power of our model—Model 3 is able to account for 26% of variation in life satisfaction. Economic factors remain statistically significant, while the newly introduced social variables are also significantly and positively correlated with life satisfaction. We find that higher levels of satisfaction with old-age and health insurance, higher (subjectively evaluated) income equity, and favorable views of government housing and healthcare policies

Table 6.3 Regression model for factors underlying urban quality of life

	Model 1	Model 2	Model 3	Model 4	Model 5	Model 6
Economic factors: absolute and relative income						
Household income (ln)	0.429 *** (0.031)	0.210 *** (0.030)	0.296 *** (0.035)	0.300 *** (0.034)	0.246 *** (0.034)	0.182 *** (0.036)
Change in living standards		0.383 *** (0.030)	0.266 *** (0.033)	0.244 *** (0.033)	0.222 *** (0.032)	0.231 *** (0.032)
Relative socioeconomic status in locality of residence (subjective rating)		0.518 *** (0.032)	0.409 *** (0.036)	0.370 *** (0.036)	0.321 *** (0.036)	0.289 *** (0.036)
Social factors: social security, social equity and public services						
Satisfaction with old-age insurance			0.104 *** (0.017)	0.092 *** (0.016)	0.084 *** (0.016)	0.080 *** (0.016)
Satisfaction with health insurance			0.064 *** (0.018)	0.050 *** (0.017)	0.047 *** (0.017)	0.047 *** (0.017)
Equality of Income Distribution			0.186 *** (0.040)	0.149 *** (0.039)	0.138 *** (0.039)	0.153 *** (0.040)
Appraisal of government housing security policies			0.226 *** (0.040)	0.182 *** (0.038)	0.175 *** (0.038)	0.159 *** (0.038)
Appraisal of government healthcare policies			0.142 *** (0.049)	0.072 (0.048)	0.057 (0.048)	0.085 * (0.048)
Environmental factors: appraisal of environmental quality						
Satisfaction with environmental quality in locality of residence				0.177 *** (0.016)	0.173 *** (0.016)	0.172 *** (0.016)
Individual quality of life: health and employment						
Economic burden due to high medical expenses					−0.340 *** (0.068)	−0.342 *** (0.068)
Economic burden due to a family member's unemployment or job instability					−0.321 *** (0.067)	−0.269 *** (0.067)
Control variables						

(continued)

Table 6.3 (continued)

	Model 1	Model 2	Model 3	Model 4	Model 5	Model 6
Female						0.137 *** (0.052)
Age (reference group: persons over 60)						
18–25						−0.104 (0.126)
26–35						−0.453 *** (0.102)
36–45						−0.568 *** (0.094)
46–60						−0.338 *** (0.088)
Married						0.315 *** (0.084)
Educational attainment (reference group: primary education and below)						
Junior secondary						0.150 * (0.090)
Upper secondary (high school/secondary vocational education)						0.279 *** (0.100)
Higher vocational education						0.465 *** (0.113)
Bachelor's degree and above						0.639 *** (0.117)
Urban household registration						
Constants	2.187 *** (0.337)	1.915 *** (0.321)	−0.720 * (0.400)	−1.144 *** (0.389)	0.001 (0.409)	0.433 (0.425)
Sample size	5162	5066	3552	3551	3530	3513
Adjusted R^2	0.044	0.170	0.261	0.295	0.307	0.325

Robust standard errors for each coefficient are included in parentheses
*** $p < 0.01$; * $p < 0.1$

all lead to greater levels of life satisfaction. This indicates that social factors have a strong influence on the life satisfaction of urban residents.

Model 4 introduces variables that measure environmental factors. We observe that life satisfaction is significantly and positively correlated with levels of satisfaction with local environmental quality (i.e. near locality of residence). The introduction of environmental factors raises the explanatory power of our model, which can now explain 29.5% of variation in life satisfaction. Clearly, environmental factors have a strong impact on residents' life satisfaction. However, after the introduction of these factors, subjective appraisals of government healthcare policies become a statistically insignificant to levels of life satisfaction.

In Model 5, factors that impact individual quality of life are included, which are measured through health and employment pressures faced by both respondents and their households. Both medical expenses and household unemployment are negatively and significantly correlated with life satisfaction. Model 5 can explain 30.7% of variation in life satisfaction. With the exception of subjective appraisals of government healthcare policies, all variables considered in Model 5 are statistically significant.

Therefore, we see that a model that incorporates both economic and non-economic factors (i.e. social and environmental factors, which are factors of the external environment, and individual quality of life) can account for nearly 31% of variation in life satisfaction, while the explanatory power of the model increases as more variables are included. This indicates that each of these variables have a strong influence on the life satisfaction of urban residents. The results of our model also indicate that our analytical framework is sufficiently robust.

All explanatory variables are included in Model 6, along with relevant control variables. Model 6 explain 32.5% of variation in life satisfaction. Subjective appraisal of government healthcare policies is statistically significant at the 10% level, while other explanatory variables are significant at the 1% level. Among the control variables, gender, marriage status, and level of educational attainment are statistically significant. Females, married respondents, and those in long-term relationships all show a higher level of life satisfaction. Similarly, those who have graduated from junior secondary school, high school, vocational education or university have higher levels of life satisfaction than those who have not received formal education or have only completed primary school. These findings are consistent with existing research (Guan 2010; Liu et al. 2012).

Interestingly, age has a U-shaped relationship with life satisfaction. Life satisfaction is greatest for the youngest group of respondents (18–25) and the oldest group (above 60). The groups in between (26–35; 36–45; 46–60) report significantly lower levels of life satisfaction, possibly because they face pressures related to work, family, and supporting elderly parents. Relevant regression coefficients are highest for those between 36 and 45, meaning that life satisfaction is lowest in this age group. This finding is also consistent with that by Appleton and Song. However, household registration is not a statistically significant factor. One obvious interpretation is that all sample data in our analysis are provided by residents of urban localities. Regardless of whether a respondent has an urban or rural household registration, they are generally entitled to use of urban infrastructure and resources, even if some public services may not be available to the latter. This lowers the impact of household registration on life satisfaction in our model.

Our analysis shows that life satisfaction in China continues to be strongly influenced by economic factors. Absolute income exerts a significant impact, but the impact thereof weakens after income levels reach a certain stage. Suppose there are two urban residents, one with an annual household income of CNY 30,000 and the other with an annual household income of CNY 80,000. Based on Model 6, life satisfaction of the wealthier resident will be higher by 0.179 points, after controlling for other factors. This gap narrows, however, as income levels increase. Life satisfaction would only be 0.088 points higher for a resident with CNY 130,000 in total annual household income than a resident with CNY 80,000. The significance of absolute income to the life satisfaction of Chinese residents is typical to the context of a developing country with high economic growth rates. This also supports the idea proposed by Diener and Biswas-Diener that the degree of social affluence is a moderating variable between income and life satisfaction. Similarly, within a specific society, general income levels moderate the effect of income on life satisfaction, which is why the income effect is weaker among high-income groups.

In our analysis, relative income has a stronger influence than absolute income on life satisfaction, which is consistent with the conclusions of existing Chinese empirical studies (Luo 2009; Guan 2010; Liu et al. 2012). This also suggests that the theoretical framework proposed by Easterlin can account for variation in life satisfaction. Among the two factors we used to measure relative income, self-assessment of one's socio-economic status relative to one's peers has a larger regression coefficient, indicating that a sense of relative superiority can significantly increase life satisfaction. In Model 6, after controlling for other variables, life satisfaction scores increase for every step up in our five-tier socio-economic ranking system. For instance, life satisfaction levels are 0.289 higher for those in the middle tier than those in the lower-middle tier. In contrast, an increase in current income of one tier over past income, for instance from "no change" to "slight change", leads to a rise of only 0.231 in life satisfaction.

Our analysis has also highlighted the impact of social and environmental factors on life satisfaction. After introducing relevant variables, the explanatory power of our model increased by nearly 13%. Among factors related to social security, satisfaction with old-age insurance had a larger regression coefficient than satisfaction with health insurance. This indicates that the former has a larger impact on overall life satisfaction, suggesting that financial security during retirement is a pertinent livelihood security issue for urban residents. Opinions on income equity also have a significant impact on life satisfaction. Suppose Resident A believes that income distribution is highly unfair, while Resident B believes that it is relatively fair. All other factors being equal, life satisfaction would be 0.306 points higher for Resident B. In addition, one's appraisal of government housing policies has the greatest impact on life satisfaction among all the indicators measuring the quality of public services.

As environmental issues, for instance haze and polluted tap water, become more widespread, public attention has turned towards environmental quality. Our model indicates that satisfaction with the living environment has a significant impact on life satisfaction. All other factors being equal, life satisfaction for a resident who is highly satisfied with their local environment (for instance, a score of 9 points) would be 1.204 points higher than a resident who is distressed with the local environmental quality (for instance, a score of 2 points).

Table 6.4 Decomposition model for environmental satisfaction (linear regression model)

Independent variable	Mean	Regression model	Independent variable	Mean	Regression model
Noise pollution	2.703 (0.926)	0.266 *** (0.029)	Poor local sanitation	2.767 (0.817)	0.915 *** (0.032)
Air pollution	2.652 (0.908)	0.359 *** (0.032)	Constants		1.426 *** (0.091)
Water pollution	2.824 (0.907)	0.222 *** (0.029)	Sample size	5407	5407
			Adjusted R^2		0.365

Note Standard deviation is included in parentheses under mean values; robust standard errors for relevant coefficients are included in parentheses under regression model values
*** $p < 0.01$

Based on the survey data, our chapter takes a closer look at residents' satisfaction with local environmental conditions. Before asking respondents to rate their satisfaction with environmental quality, the CGSS 2013 asked them to rate environmental issues on a four-point scale, including noise pollution, air pollution, water pollution, and the cleanliness of their local residential area. A score of 1 meant that the issue was "very severe", 2 meant "quite severe", 3 meant "not very severe", while 4 meant that the issue was non-existent. As shown in Table 6.4, air pollution was reported as the most serious issue (average score of 2.652), with 39% of respondents considering it to be "very severe" or "quite severe". This is closely followed by noise pollution, with an average score of 2.703.

Our chapter uses four environmental pollution indicators as independent variables to construct a regression model for environmental satisfaction. The results of our model show that these four selected indicators each have a significant impact, allowing our model to explain 36.5% of variation in environmental satisfaction. Residents' appraisal of the cleanliness of their local environment turned out to have the strongest impact. After controlling for other variables, a belief that local sanitation is not a severe issue increases environmental satisfaction levels by 1.83 points. This is followed by appraisals of air quality, where environmental satisfaction increases by 0.718 points for residents who believe that air pollution is "not very severe". Our empirical analysis underscores the importance of regular urban sanitation and air quality in improving the happiness and life satisfaction of residents.

6.5 Conclusion and Discussion

Our chapter has constructed a framework for analyzing the factors underlying life satisfaction before conducting an empirical analysis of the life satisfaction of urban residents. We first proposed that a framework for analyzing life satisfaction should include economic, social, environmental and individual quality of life factors. Such a framework would incorporate the material conditions necessary to sustain daily life, as well as variables related to the external environment and the pressures of life faced by an individual, thus allow us to more comprehensively survey the factors that affect life satisfaction.

Empirical analysis then allows us to understand trends of change in the life satisfaction of urban residents, while also verifying the validity and robustness of our analytical framework. Our study showed that the average urban resident was quite satisfied with life (average score of 6.84). Further empirical analysis then reveals that both absolute and relative income had a significant impact on life satisfaction. Among the two variables for relative income, income levels compared with one's peers had a stronger impact.

Our analysis has also incorporated social and environmental factors. Social factors are measured through the level of social security, level of social equity, and quality of public services, each of which have a significant impact on the life satisfaction of urban residents. Meanwhile, the greater public attention on environmental issues is reflected in our analysis, which shows that greater satisfaction with local environmental conditions is significantly correlated with greater overall life satisfaction. A more detailed analysis then found that, out of four environmental indicators, levels of local sanitation and air quality have the greatest impact on environmental satisfaction. Individual quality-of-life factors, as measured by pressures arising from medical expenses and family members' unemployment are significantly and negatively correlated with life satisfaction.

Our analysis also found that, despite the importance of income to life satisfaction, money alone does not lead to happiness. The social security system, income distribution, and environmental quality are significantly correlated with life satisfaction. As China's economy continues to develop, people will come to demand a greater quality of life after their basic needs are fully met. Such demands will include safe food products, clean air, quality education, and a clean government. These demands require the government to replace the "GDP at all costs" model with a people-oriented development model. After all, continued support by the people requires progress in relevant institutions and policies, which are necessary for sustaining upward trends in life satisfaction. Whether or not the "turning point" of the Easterlin paradox can be successfully navigated represents a major challenge for China's socio-economic development.

As Premier Li Keqiang pointed out in his government work report at the Third Session of the Twelfth National People's Congress, "developing a tightly woven and sturdy safety net to secure and improve living standards" requires "put[ting] people first".[6]

Our study finds four policy implications.

Firstly, policy should seek to sustain growth in real incomes, as well as focus on increasing the incomes of less affluent social groups, which would greatly enhance their life satisfaction.

Secondly, satisfaction with life and government policies can be raised by enhancing livelihood security. This can be achieved through improvements to existing

[6] https://language.chinadaily.com.cn/2015-03/17/content_19834004_2.htm.

social security systems, such as housing security, and old-age and health insurance, greater government efficiency, and better public services.

Thirdly, there should be greater reforms to income distribution systems. More equitable income distribution would lead to a greater sense of income equity and confidence in life.

Fourthly, environmental governance should be an important part of the government agenda. The government should focus on environmental issues of public concern, strengthen relevant enforcement efforts, and promote the development of a circular economy, in order to achieve sustainable growth in national wellbeing.

The results of our analysis also bear implications for further research on sustainable consumption. For example, Easterlin and Veenhoven both argue that when a country or individual reaches a certain level of material wealth, an increase in income does not necessarily lead to an increase in subjective wellbeing. Empirical analysis has also shown that material conditions are not the most important underlying factors for either happiness or life satisfaction. Therefore, sustained growth in consumption will not lead to significant increases in subjective wellbeing. In contrast, sustainable consumption practices will lead to greater subjective wellbeing through richer inner experiences and sense of accomplishment. This ensures the legitimacy of sustainable consumption practices and principles: lower consumption levels or changes in consumption patterns are desirable if they result in lower environmental harm without sacrificing subjective wellbeing.

This justifies government efforts to encourage green consumption, which will not only enhance ecological balance, but also drive economic growth by popularizing new forms of consumption. During this process, government should strive to create an external environment that encourages the adoption of green consumption practices. Government efforts in this respect should include improvements to infrastructure, enhancements to the quality and administration of public services, and stronger environmental governance, which also happen to have a strong influence on subjective wellbeing. Research has also shown that continual satisfaction of ever greater material desires does not lead to happiness. Instead of lifestyles that emphasize excessive, conspicuous consumption, people should focus more on the quality of life, including a sound work-life balance, good psychological and physical health, and enjoying the sense of achievement that comes from work.

Of course, our research is still lacking in many respects. Because of the limited range of our data, our model does not include certain relevant objective indicators, as well as data on regarding how respondents spend their time. Our model may also be extended in future research, both horizontally and vertically. With regards to a "vertical" extension, panel data or multi-year cross-sectional data can be used to investigate if socio-economic development and changes to demographic/income structures have led to changes in the effects or mechanisms of economic, social, environmental and individual quality of life factors. With regards to "horizontal" extension, one may investigate if the impact of such factors is affected by reason of gender, age, or level of educational attainment. For instance, life satisfaction for the more highly educated may be less affected by economic factors but are more influenced by the level of social equity.

References

Bai, Weichun, and Demen Shao. 1994. Taolun zhengfu pingjia wenti [On Evaluating Government Performance]. *Journal of Northeast Normal University (Philosophy and Social Sciences Edition)* (5).

Bao, Zonghua. 2003. Guanyu fangjia shourubi de zaiyanjiu [A New Look at Housing Price-to-Income Ratios]. *Urban Development* (1).

Baudrillard, Jean. 2001. *La Société de consommation*. Translated by Chengfu Liu and Zhigang Quan as *Xiaofei shehui* [Consumer society]. Nanjing: Nanjing University Press.

Becker, Gary. 2007. *Human Capital: A Theoretical and Empirical Analysis with Special Reference to Education*. Translated by Hong Guo. Beijing: China Citic Press.

Bian, Yanjie, and Yongli Liu. 2005. Social Stratification, Home Ownership and Quality of Living: Evidence from China's Fifth Census. *Sociological Studies* (3). (in Chinese).

Bian, Yanjie, and Yang Xiao. 2014. A Comparative Analysis of Subjective Well-Being in China and Britain. *Sociological Studies* (2). (in Chinese).

Bian, Yanjie, Logan John, Hanlong Lu, Yunkang Pan, and Ying Guan. 1996. Danweizhi yu zhufang shangpinhua [Work Units and the Commodification of Housing]. *Sociological Studies* (1).

Blau, Peter. *Inequality and Heterogeneity*. Translated by Chunguang Wang and Shengzan Xie. Beijing: China Social Sciences Press.

Cai, He, and Jianhong Huang. 2013. Shei yongyou diertao fang? -shichang zhuanxing yu chengshi zhufang fenhua [Who Owns a Second Home? Market Transformation and Urban Housing Stratification]. *Jilin University Journal Social Sciences Edition* (4).

Cai, Fang, Yang Dou, and Meiyan Wang. 2001. Household Registration System and Labor Market Protection. *Economic Research Journal* (12). (in Chinese).

Cai, Fang. Weishenme laodongli liudong weiyou suoxiao chengxiang shouru chaju? [Why Has Labor Mobility Not Narrowed the Urban-Rural Income Gap?]. *Theory Front* (20).

Chen, Wanting, and Xiumei Zhang. 2013. Woguo jumin zhuguan xingfugan ji qi yingxiang yinsu fenxi – jiyu CGSS 2010 nian shuju" [Analysis of Chinese Residents' Subjective Level of Happiness and the Influencing Factors-Based on Data from the CGSS 2010]. *The World of Survey and Research* (10).

Chen, Zongsheng and Yunbo Zhou. 2001. The Impact of Illegal and Abnormal Incomes on Income Inequality of Inhabitants and Its Economic Analysis. *Economic Research Journal* (4). (in Chinese).

Chen, Yao. 1998. Zhengzhi pingjia lun [Political Appraisal Theory]. *Social Sciences in Nanjing* (1).

Cao, Dayu. 2011. Huanjing zhiliang yu jumin shenghuo manyidu de shizheng fenxi [Empirical Analysis of Environmental Quality and Residents' Life Satisfaction. *Statistics and Decision* (21).

Deng, Quheng. 2017. Earnings Differential Between Urban Residents and Rural Migrants: Evidence from Oaxaca-Blinder and Quantile Regression Decompositions. *Chinese Journal of Population Sciences* (2). (in Chinese).

Fang, Gang, and Xiaotian Feng. 2009. Research on Subjective Well-Being Differences between Urban and Rural Residents and Their Influence Factors—Take Chengdu City as a Case. *Population and Development* (6). (in Chinese).

Farrell, Henry. "The Consequences of the Internet for Politics". Translated and edited by Zheng Ying and Li Li. *Foreign Theoretical Trends*, no. 1 (2013).

Feng, Xiaotian, and Songguo Yi. 2000. Chengshi jumin jiating shenghuo zhiliang: zhibiao ji qi jiegou [Quality of Urban Residents' Family Life: Indicators and Their Structure]. *Sociological Studies* (4).

Feng, Xiaotian. 2009. Shenghuo zhiliang yanjiu: Jin sanshi nian huigu ji xiangguan wenti tantao [A Study on Quality of Life: Review of Achievements over the Past 30 years and Discussion on Related Issues]. *Social Science Research* (6).

Foner, Eric. 2010. *Give Me Liberty! An American History (vol. 2)*. Translated by Xi Wang. Beijing: Commercial Press.

Fu, Hao, and Yimin Wang. 2014. Daxuesheng zhuguan xingfugan xianzhuang diaocha yanjiu [A Survey of the Subjective Well-Being of College Students]. *Journal of Agricultural University of Hebei (Agriculture and Forestry Education Edition)* (4).
Fukuyama, Francis. *Trust: The Social Virtues and the Creation of Prosperity*. Translated by Zhihua Peng. Haikou: Hainan Publishing House.
Galbraith, John Kenneth. 1965. *The Affluent Society*. Translated by Shiping Xu. Shanghai: Shanghai People's Publishing House.
Gao, Hongying, and Yuanjiang Miao. 2006. The Develop Locus of Well-Being Study in the West. *Journal of Jinggangshan University (Natural Science)* (4). (in Chinese).
Gao, Jinjin. 2007. Huanbao NGO ruhe tisheng zishen fazhan nengli [How Environmental NGOs Enhance Their Development Capabilities]. *Greenleaf* (5).
Georg, Simmel. 2001. *Philosophie der Mode* [The Philosophy of Fashion]. Translated by Yong Fei as *Shishang de zhexue*. Beijing: Culture and Art Publishing House.
Guan, Hao. 2010. Shouru dui xingfugan de yingxiang yanjiu: juedui shuiping he xiangdui diwei [The Impact of Income on Happiness: Absolute and Relative Income]. *Nankai Economic Studies* (5).
Guo, Xiaoyan. 2014. Analysis on Rural Residents' Health Seeking Behavior and Its Influencing Factors—Taking Chaohu City, Anhui Province as an Example. *Journal of Anhui Agricultural Sciences* (33). (in Chinese).
Guojia Fagaiwei Shehui Fazhan Yanjiusuo Ketizu [Research Group from the Institute for Social Development, China National Development and Reform Commission]. Kuoda zhongdeng shouruzhe bizhong de shizheng fenxi he zhengce jianyi [Empirical Evidence and Policy Proposals for Raising the Proportion of the Middle-Income Group]. *Economics Perspectives* (5).
Guojia Tongjiju Chengdiao Zongdui Ketizu [Research Group from the National Bureau of Statistics Urban Survey Team]. "60,000–500,000 yuan: Zhongguo chengshi zhongdeng shouru qunti tanjiu [CNY 60,000–500,000: An Inquiry into China's Urban Middle-Income Groups]. *Data*, no. 6 (2005).
Han, Jun (ed.). 2009. *Strategic Research on China's Migrant Workers*. Shanghai: Shanghai Far East Publishers. (in Chinese).
Hu, Hongshuhao, and Yuanping Lu. 2012. Income Inequality, Health and Subjective Well-Being Among the Elderly: Empirical Evidence on the Aging of China. *China Soft Science* (11). (in Chinese).
Hu, Rong. 2011. Housing Tenure Choice from Regional Disparity Perspective: Evidence from the Analysis of CGSS 2006. *Journal of Guangzhou University (Social Science Edition)* (12). (in Chinese).
Hu, Rong. 2012. Housing Inequality During the Market Transition: Evidence from the Data of CGSS 2006. *Chinese Journal of Sociology* (1). (in Chinese).
Jin, Chengwu. 2009. Income Difference between Employees with Urban and Rural Hukou in the Urban Labor Market. *Chinese Journal of Population Science* (4). (in Chinese).
Jing, Qiao. 2012. Rethinking Political Participation in the View of Social Quality Theory: Also About Foreign Concepts' Localization. *Journal of Social Sciences* (8). (in Chinese).
Jung, Joon-ho. 2004. Fei zhengfu zuzhi de zhengce canyu ji yingxiang moshi [Policy Participation and Impact of Non-governmental Organizations]. *Chinese Public Administration* (5).
Kalton, Graham. 2014. *Introduction to Survey Sampling*. Shanghai: Truth & Wisdom Press.
Lai, Desheng. Lun laodongli shichang de zhiduxing fenge [On Structural Division in the Labor Market]. *Economic Science* (6).
Li, Yue, and Hongzhi Cui. 2014. Nongcun laoren zhuguan xingfugan ji qi yingxiang yinsu fenxi – jiyu Shandong, Henan, Shaanxi sansheng nonghu diaocha shuju fenxi [An Analysis of the Subjective Well-Being of Elderly Residents in Rural Areas and Underlying Factors: Based on Data from Surveys of Rural Residents in Shandong, Henan and Shaanxi provinces]. *China Rural Survey* (4).
Li, Shi, and Sai Ding. 2003. Long-Term Change in Private Returns to Education in Urban China. *Social Sciences in China* (6). (in Chinese).

Li, Shi, and Wenbin Li. 1994. Zhongguo jiaoyu touzi de geren shouyilü de guji [Estimates of Individual Returns on Investment in Education in China]. In *The Distribution of Income in China*, ed. Renwei Zhao and Griffin Keith. Beijing: China Social Sciences Press. (in Chinese).

Li, Peilin, and Feng Tian. 2014. The New Generation of Migrant Workers: Social Attitudes and Behavioral Choices. *Chinese Journal of Sociology* (3). (in Chinese).

Li, Bin, and Kai Wang. 2010. Zhongguo shehui fenceng yanjiu de xin shijiao – chengshi zhufang quanli de zhuanyi [A New Perspective in Research on Social Stratification in China—Transference of Urban Housing Rights]. *Exploration and Free Views* (4).

Li, Deming, Tianyong Chen, and Zhenyun Wu. 2007. Quality of Life and Subjective Well-Being of the Aged in Chinese Countryside. *Chinese Journal of Gerontology* (12). (in Chinese).

Li, Chunling. 2012. 80 hou daxue biyesheng jiuye zhuangkuang ji yingxiang yinsu fenxi – jiyu 6 suo 985 gaoxiao biyesheng de diaocha [An Analysis of the Employment Status of University Graduates Born in the 1980s—Based on a Survey of Graduates from 6 "Project 985" Universities]. *Jiangsu Social Sciences* (3).

Li, Shi. 1997. Zhongguo jingji zhuangui zhong laodongli liudong moxing [Models of Labor Mobility During China's Economic Transition]. *Economic Research Journal* (1).

Li, Peilin. 2004. *Social Stratification in China Today*. Beijing: Social Sciences Academic Press. (in Chinese).

Li, Peilin. 2007. Kuoda zhongdeng shouruzhe bizhong de duice silu [Strategical Approaches for Increasing the Proportionate Size of the Middle-Income Group]. *Chinese Journal of Population Science* (5).

Li, Bin. 2009. *Housing Polices in Differentiation: An Evaluation Study on Urban Housing Reforms*. Beijing: Social Sciences Academic Press. (in Chinese)

Li, Chunling. 2011. Zhongchan jieji de xiaofei shuiping he xiaofei fangshi [Level and Patterns of Consumption by the Middle Class]. *Social Sciences in Guangdong* (4).

Li, Shi. 2012. Lixing panduan woguo shouru chaju de bianhua qushi [Rationally Determining the Changing Trends in China's Income Gap]. *Exploration and Free Views* (8).

Li, Peilin. 2014. Shehui zhili yu shehui tizhi gaige [Social Governance and Reform of the Social System]. *Journal of Chinese Academy of Governance* (4).

Liang, Zhongming. 1999. Guanyu zhengfu pingjia wenti de jidian sikao [Some Thoughts on the Evaluation of Government Performance]. *Journal of Shaanxi Administration School and Shaanxi Economic Management School* (1).

Lin, Zhili. 2002. Shengtai huanjing wenti dui zhengzhi huodong de yingxiang [The Impact of Eco-Environmental Issues on Political Activities]. *Journal of Chongqing Normal University (Edition of Social Sciences)* (3).

Liu, Shishun, and Wenquan Ling. 2009. Multiple Mediation Models and Their Applications. *Psychological Science* (2). (in Chinese).

Liu, Junqiang, Moulin Xiong, and Yang Su. 2012. National Sense of Happiness in the Economic Growth Period: A Study Based on CGSS Data. *Social Sciences in China* (12). (in Chinese)

Liu, Suhua. 2005. Jiuye zhiliang: neihan ji qi yu jiuye shuliang de guanxi [Employment Quality: Implications and Relationship with Employment Rates]. *Inner Mongolia Social Sciences* (5).

Liu, Jingming. 2006. The Change of Labor Market Structure and Returns on Human Capital. *Sociological Studies* (6). (in Chinese).

Liu, Xiangjun. 2010. Shi lun gongmin pingjia zhengfu de tujing [On Channels for Citizens to Evaluate Government Performance]. *Tribune of Social Sciences* (8).

Lu, Chuntian, Dayong Hong, and Gong Cheng. 2014. Dui chengshi jumin pingjia zhengfu huanbao gongzuo de zonghe fenxi – jiyu CGSS 2003 he CGSS 2010 shuju [A Comprehensive Analysis of Urban Residents' Appraisal of Government Efforts in Environmental Protection-Based on Data from the CGSS 2003 and CGSS 2010]. *Theoretical Exploration* (2).

Lu, Xueyi (ed.). 2002. Dangdai Zhongguo shehui jieceng yanjiu baogao [Research Report on Social Stratification in Contemporary China]. Beijing: Social Sciences Academic Press.

Lu, Xueyi. 2003. Nongmingong tizhi xuyao genben gaige [Fundamental Reforms to China's Migrant Worker System Are Required]. *China Reform (Rural Edition)* (12).

Lu, Yulong. 2005. Gongyou chanquan: Jingji shiyong fang zhidu chuangxin yanjiu [Common Property Rights: A Study on Innovations in Systems for Affordable Housing]. *Real Estate Information of China* (9).
Lu, Sicheng. 2011. Gaishan NGO de shengcun tiaojian shi bu ke huibi de xianshi [It Is a Fact that Conditions for NGO Survival Must Be Improved]. *Greenleaf* (3).
Luo, Chuliang. 2009. Absolute Income, Relative Income and Subjective Well-being: Empirical Test Based on the Sample Data of Urban and Rural Households in China. *Journal of Finance and Economics* (11). (in Chinese).
Ma, Guanghong, and Yilin Tian. 2010. Zhongguo gonggong zhufang lilun yu shijian yanjiu [A Study of the Theory and Practice of Public Housing in China]. Beijing: China Architecture & Building Press.
Ma, Deyong. 2008. Trust, Its Origin and Transition in Mainland China. *Open Times* (4). (in Chinese).
Ma, Liang. Chengxiang jumin zhuguan xingfugan bijiao yu xingfu gansu jianshe yanjiu [A Comparison of Subjective Well-Being Between Urban and Rural Households and Research on the Building of a "happy gansu"]. *China-Today Forum* (17).
Mead, Margaret. 1988. *Culture and Commitment: A Study of the Generation Gap*. Translated by Hu Zeng. Beijing: Guangming Daily Publishing House.
Miao, Yuanjiang. 2007. The Wellbeing Concept Model Evolve. *Journal of Gannan Normal University* (4). (in Chinese).
National Bureau of Statistics. 2013. *China Statistical Yearbook 2013*. An English version is available from http://www.stats.gov.cn/english/Statisticaldata/AnnualData/
Peng, Siqing. 1999. Xinren de jianli jizhi: guanxi yunzuo yu fazhi shouduan [Mechanisms for Establishing Trust: The Function of *guanxi* and the Role of the Legal System]. *Sociological Studies* (2).
Qi, Lingling and Jingping Lai. 2013. Zhengfu xingwei, zhengzhi xinren yu zhuguan xingfugan [Government Actions, Political Trust and Subjective Wellbeing]. *Academic Research* (7).
Qiao, Mingrui, Xueya Qian, and Xianguo Yao. 2009. Labor Market Segmentation, *Hukou* and Urban-Rural Difference in Employment. *Chinese Journal of Population Science* (1). (in Chinese).
Qin, Qiong. 2014. Laonianren zhuguan xingfugan xingbie chayi de yuanfenxi [A Meta-Analysis on Gender Differences in the Subjective Wellbeing of the Elderly]. *Ability and Wisdom* (27).
Rostow, W.W. 2001. *The Stages of Economic Growth*. Translated by Xibao Guo and Songmao Wang. Beijing: China Social Sciences Press.
Shi, Yaojiang, and Yu Cui. 2006. Gongmin gongpingguan ji qi dui shehui gongping pingjia he shenghuo manyidu yingxiang fenxi [Citizens' Views on Fairness and Analysis of the Impact on their Life Satisfaction and Assessment of Social Equity]. *Management World* (10).
Simmel, Georg. 2001a. *Philosophie der Mode*. Translated by Yong Fei as *Shishang de zhexue* [The Philosophy of Fashion]. Beijing: Culture and Art Publishing House.
Su, Shishang. 2007. Jiuye zhiliang wenti yanjiu – guoneiwai jiuye zhiliang yanjiu de lilun he zhengce zongshu [A Study on the Quality of Employment: A Review of Theoretical Research and Relevant Policies in China and Abroad]. Master's thesis, Capital University of Economics and Business.
Sun, Xiaofei, and Sanguo Tian. 2010. The Research on College Students' Subjective Well-Being Based on Different Sexes. *Journal of Hebei University of Engineering (Social Science Edition)* (3). (in Chinese).
Sun, Xin, Zhigang Xu, Ran Tao and Fubing Sun. 2007. Impacts of Political Trust and Social Capital on Villagers' Committee Election Participation in China. *Sociological Studies* (4). (in Chinese).
Sun, Wenkai, Chongen Bai and Peichu Xie. 2011. The Effect on Rural Labor Mobility from Registration System Reform in China. *Economic Research Journal* (1). (in Chinese).
Sun, Lin, Aiqin Zhang, Xinyan Chu, Jie Chen and Qingzhong Pan. 2012. Nongcun jumin zhuguan xingfugan yingxiang yinsu fenxi [An Analysis of the Factors Underlying Rural Residents' Subjective Well-Being]. *Jiangsu Agricultural Sciences* (3).

Sun, Chunmei, Ning Zhang, and Jin Tian. 2014. Survey of the Health Seeking Behavior of the Residents of a Community in Shanghai. *Shanghai Medical & Pharmaceutical Journal* (18). (in Chinese).

Sun, Hui. 2012. Empirical Research of the Structural Transformation of Urban Residents' Income. *Statistical Research* (10). (in Chinese).

Sylvie, Démurger, Gurgand Marc, et al. 2009. Migrants as Second-Class Workers in Urban China? A Decomposition Analysis. Translated by Guicang Li. *Foreign Theoretical Trends* (8).

Tang, Dan, Zou Jun, Shen Jiliang and Zhang Ling. 2006. The Influence Factors of Subjective Well-being in Older Adults. *Chinese Mental Health Journal* (3). (in Chinese).

Tang, Minghao, Qing Zhou and Haimin Kuang. 2009. Chengzhen jumin huanjing taidu yu huanjing xingwei de diaocha [A Survey of Urban Residents' Environmental Attitudes and Behavior]. *Journal of Xiangtan Normal University (Natural Science Edition)* (1).

Tang, Lijian. 2014. Xinxing nongcun hezuo yiliao zhidu xia nongucn jumin menzhen jiuyi xingwei fenxi [Analysis of Rural Residents' Healthcare Seeking Behavior Under the New Rural Cooperative Medical Scheme]. *Soft Science of Health* (12).

Tian, Feng. 2009. Children Growing up During the Reform and Opening up: The Post-70 and Post-80's Consciousness of Fairness and Democracy in China. *Youth Studies* (6). (in Chinese).

Tian, Feng. 2010. A Study of the Income Gap between Urban Workers and Migrant Workers. *Sociological Studies* (2). (in Chinese).

Tong, Zhifeng. 2008. Kuaisu zhuanxingqi xia de huanjing quntixing shijian yanjiu [A Study of Environmental Mass Incidents in an Era of Rapid Transition]. In *Zhixu yu jinbu: Zhejiang xiangcun shehui jubian licheng yu jingyan lilun yantaohui lunwen ji* [Order and Progress: Collection of Papers from a Conference on Social Change in Rural Zhejiang and Relevant Empirical Theory].

Tong, Zhifeng. 2014. On Environmental Organizational Changes and Environmental Development in Social Movement Network: An Analysis to Environmental Protest in P County, Fujian Province. *Journal of Nanjing University of Technology* (1). (in Chinese).

U.S. Department of Health and Human Services, November 15, 2013. http://aspe.hhs.gov/poverty/13poverty.cfm

Wang, Shuming, and Xiangfeng Yang. 2011. The Overview of Research on National Environmental Organizations. *Journal of Social Science of Jiamusi University* (6). (in Chinese).

Wang, Junxiu, and Yiyin Yang. 2011. *Annual Report on Social Mentality of China*. Beijing: Social Sciences Academic Press. (in Chinese).

Wang, Jun, Zhongze Chang, and Hong Liu. 2008. Health Care Demand and Choice of Medical Treatment. *Economic Research Journal* (7). (in Chinese).

Wang, Meiyan. 2003. Zhuangui shiqi de gongzi chayi: qishi de jiliang fenxi [Wage Differences During a Period of Transition: A Quantitative Analysis of Discrimination]. *The Journal of Quantitative and Technical Economics* (5).

Wang, Meiyan. 2005. Employment Opportunities and Wage Gaps in the Urban Labor Market: A Study of the Employment and Wages of Migrant Laborers. *Social Sciences in China* (5). (in Chinese).

Wang, Xiaolu. 2010. Woguo guomin shouru fenpei xianzhuang, wenti ji duice [The Present State, Problems and Solutions of National Income Distribution. *Journal of Chinese Academy of Governance* (3).

Wang, Fuqin. 2010. Human Capital, Labor Market Segmentation and Income Distribution. *Chinese Journal of Sociology* (1). (in Chinese).

Wang, Junxiu. 2011. OECD de xingfu zhishu ji dui woguo de jiejian yiyi [The OECD Better Life Index and Its Significance as a Reference for China]. *Democracy & Science* (6).

Weber, Max. 2008. *Konfuzianismus und Taoismus* [Confucianism and Taoism]. Nanjing: Jiangsu People's Publishing House.

Wei, Mingkai, and Baoan Fang. 2006. Woguo jiuye zhiliang zhuangkuang fenxi [Analysis of the State of Employment Quality in China]. *Industrial & Science Tribune* (12).

Wei, Min, Jincheng Xiao, Shanfa Yang, and Zhen Lü. 2014. Bibliometric Analysis on the Influencity Factors Related to Health Care Seeking Behavior of Rural Residents from 2003–2013. *Medicine and Society* (3). (in Chinese).
Wen, Xiaoliang, Jian Mi, and Lizhi Zhu. 2011. Subjective Well-Being and Its Factors in China from 1990 to 2007. *Finance and Trade Research* (3). (in Chinese).
Wen, Zhonglin, Lei Zhang, Jietai Hou, and Hongyun Liu. 2004. Testing and Application of the Mediating Effects. *Acta Psychologica Sinica* (5). (in Chinese).
Wright, Mills C. 2006. *White Collar: The American Middle Classes.* Translated by Xiaohong Zhou. Nanjing: Nanjing University Press.
Wu, Jiannan, and Qiushuang Zhuang. 2004. 'Zishangerxia' pingjia zhengfu jixiao tansuo: 'gongmin pingyi zhengfu' de deshi fenxi" [Bottom-Up Assessment of Government Performance: The Advantages and Disadvantages of "Citizen Appraisal of Government Performance]. *Theory and Reform* (5).
Wu, Junping. 1998. *Di wu dai ren* [The Fifth Generation]. Tianjin: Tianjin Education Press.
Wu, Xiangrong. 2009. The Research of Yield on Human Capital of Rural Migrant Workers. *Youth Studies* (4). (in Chinese).
Xiao, Yong. 2002. Standards of Evaluating a Government. *Journal of Leshan Normal University* (1). (in Chinese).
Xiao, Wentao. 2004. Huji zhidu baohu le eryuan laodongli shichang ma? [Does the Household Registration System Protect the Dual Labor Market?]. *Chinese Rural Economy* (3).
Xie, Sisheng, and Xianguo Yao. 2006. Nongmingong gongzi qishi de jiliang fenxi [A Quantitative Analysis of Wage Discrimination Against Migrant Workers]. *Chinese Rural Economy* (4).
Xie, Guihua. 2007. Migrant Workers and Urban Labor Market. *Sociological Studies* (5). (in Chinese).
Xing, Zhanjun. 2004. Chengshi jumin de zhuguan xingfugan yingxiang yinsu [The Factors Underlying Urban Residents' Subjective Wellbeing. *The New Orient* (11).
Xing, Zhanjun. 2008. The Development of the Revised Subjective Well-Being Scale for Chinese Citizens (SWBS-CC) and Its Nationwide Norms. *Psychological Science* (6). (in Chinese).
Xing, Zhanjun. 2011. A Study of the Relationship between Income and Subjective Well-Being in China. *Sociological Studies* (1). (in Chinese).
Xiu, Yan, Biao Xu, Chengren Zhu, Weihua Gao, Benfu Yang, and Qingwu Jiang. 2003. Xingbie yinsu dui nongcun jiehebing huanzhe jiuyi xingwei de yingxiang yanjiu [A Study on the Influence of Gender in the Healthcare Seeking Behavior of Rural Tubercular Patients]. *China Public Health* (5).
Xu, Jianhua, Chengming Chen, and Xiang An. 2003. Dui zhongdeng shouru de jieding yanjiu [A Study on the Definition of "Middle Income"]. *Shanghai Statistics* (8).
Xu, Jialiang. 1994. Lun zhengfu pingjia [On the Evaluation of Government Performance]. *Journal of Hangzhou University (Philosophy and Social Sciences Edition)* (4).
Xu, Hao. 2008. Tonggong butong chou' yuanyu yonggong 'shuangguizhi' – zhuanfang renbaobu longdong gongzi yanjiusuo suozhang Su Hainan ["Different Wages for the Same Work" Result from the "Dual-Track System": An Interview with Su Hainan, Director of the Wage Research Institute at the Ministry of Human Resources and Social Security]. *China Economic Weekly* (25).
Yan, Biaobin, Xue Zheng, and Lin Qiu. 2003. A Study of the Factors Which Influence the Subjective Well-Being of College Students. *Journal of South China Normal University (Natural Science)* (2). (in Chinese).
Yang, Ming, Tianguang Meng, and Ran Fang. 1990. Social Trust in a Changing Society: Stock and Change from 1990 to 2010. *Journal of Peking University (Philosophy & Social Sciences)* (6). (in Chinese).
Yang, Yanchun, Mutao He, Changming Zhu, and Yugen Ma. 1998. Laoren xingfudu yu shehui xinli yinsu de diaocha yanjiu [A Survey of the Relationship Between Socio-psychological Factors and Well-Being Among the Elderly]. *Chinese Mental Health Journal* (1).
Yang, Zhongfang. 1994. Zhonggguoren zhende shi jiti zhuyi de ma? – Shilun Zhongguo wenhua de jiazhi tixi [Are the Chinese Really Collectivist? On the Value Systems of Chinese Culture].

In *Chinese Values from the Viewpoint of Social Sciences.* Taipei: Guiguan Tushu Gongsi. (in Chinese).

Yang, Yiyin. 1998. Shehui xinli lingyu de jiazhiguan yanjiu shuyao [A Summary of Research in the Field Of Social Psychology on Value Systems]. *Social Sciences in China* (2).

Yang, Yiyong. 2001. Laodongli shichang de xingzheng fenge [Administrative Segmentation of the Labor Market]. *Review of Economic Research* (27).

Yang, Decai. 2012. Lun renli ziben eryuanxing dui chengxiang shouru chaju de yingxiang [On the Effect of the Dual Nature of Human Capital on the Urban-Rural Income Gap]. *Contemporary Economic Research* (10).

Yao, Xianguo, and Puqing Lai. 2004. Urban-rural *Huji* Differentials in Chinese Labor Relations. *Economic Research Journal* (7). (in Chinese).

Yao, Zhaoyu, and Lei Zhang. 2013. Impact of the New Rural Cooperative Medical System on Farmers' Health Seeking Behavior: A Comparative Analysis on Three Cities of South Jiangsu. *Journal of Nanjing Agricultural University (Social Sciences Edition)* (1). (in Chinese).

Yu, Hongsheng. 2001. Gongmin zhengzhi pingjia ji qi zuoyong yu yiyi [Citizens' Evaluation of Politics and Its Effect and Significance. *Literature, History and Philosophy* (5).

Zhang, Yi. 2003. Zhongguo jiecengnei hunzhi de yanxu [The Perpetuation of Class Endogamy in China]. *Chinese Journal of Population Science* (4).

Zhang, Chewei. 2006. Human Capital Return and Income Disparity: 'Matthew Effect' and Its Implication. *Economic Research Journal* (12). (in Chinese).

Zhao, Bin, and Mina Liu. 2013. Shouru, shehui ziben, jiankang yu chengshi jumin xingfugan de shizheng fenxi [An Empirical Analysis of the Relationships Between Urban Residents' Subjective Well-Being and Their Income, Social Capital and Health]. *Statistics & Decision* (20).

Zhong, Yiping, Qianbao Tan, and Ying Zhang. 2004. Daxuesheng huanjing yishi yu huanjing xingwei de diaocha yanjiu [A Survey and Study of the Environmental Consciousness and Behavior of College Students]. *Journal of Psychological Science* (3).

Zhongguo gongzhong huanjing suzhi pinggu zhibiao tixi yanjiu, xiangmu ketizu ["Research on Systems of Indicators for Public Appraisal of Environmental Quality in China" Research Group], comp. *Zhongguo gongzhong huanjing suzhi pinggu zhibiao tixi yanjiu* [A Study of Indicator Systems for Public Appraisal of Environmental Quality in China]. Beijing: China Environmental Science Press.

Zhou, Changcheng, and Na Ren. 2006. Economic Development and Subjective Quality of Life: A Case Study from Beijing, Shanghai and Guangzhou. *Wuhan University Journal (Philosophy & Social Sciences)* (2). (in Chinese).

Zhou, Jianchu, Zhenhuan He, Deqi Yang, and Mutao He. 1988. Xingfudu liangbiao zai 114 ming yi zu laoren zhong de yingyong [Application of the Memorial University of Newfoundland Scale of Happiness on 114 Elderly Members of the Yi Ethnic Group]. *Chinese Journal of Gerontology* (1).

Zhou, Yi. Daigou xianxiang de shehuixue yanjiu [A Sociological Perspective on the Generation Gap]. *Sociological Studies* (4).

Zhou, Xiaohong, (ed.). 2005. *Survey of the Chinese Middle Classes.* Beijing: Social Sciences Academic Press. (in Chinese).

Zhou, Zhijia. 2008. Environmental Protection, Group Pressure or Interests Relatedness? *Chinese Journal of Sociology* (4). (in Chinese).

Zhuang, Zixi. 2009. The Citizens' Participation in Government Performance Evaluation. Master's thesis, Shanghai Jiao Tong University. (in Chinese).

马克斯·韦伯, 2008,《儒教与道教》, 南京: 江苏人民出版社。

The manufacturer's authorised representative in the EU is Springer
Nature Customer Service Centre GmbH, Europaplatz 3, 69115 Heidelberg,
Germany. If you have any concerns regarding our products, please
contact ProductSafety@springernature.com

Printed and bound by CPI Group (UK) Ltd, Croydon, CR0 4YY
30/09/2026
02225264-0011